5G Cyber Risks and Mitigation

5G technology is the next step in the evolution of wireless communication. It offers faster speeds and more bandwidth than 4G. One of the biggest differences between 4G and 5G is that 5G will be used for a wider range of applications. This makes it ideal for applications such as autonomous vehicles, smart cities, and the Internet of Things (IoT). This means that there will be more devices connected to 5G networks, making them more vulnerable to cyber attacks. However, 5G also introduces new cyber risks that need to be addressed. In addition, 5G networks are expected to be much more complex, making them harder to secure.

5G networks will use new technologies that could make them more vulnerable to attacks. These technologies include massive multiple input, multiple output (MIMO), which uses more antennas than traditional cellular networks, and millimeter wave (mmWave), which uses higher frequencies than traditional cellular networks. These new technologies could make it easier for attackers to intercept data or disrupt service.

To address these concerns, security measures must be implemented throughout the network. Security mechanisms must be included in the design of 5G networks and must be updated as new threats are identified.

Moreover, to address these risks, 5G security standards need to be developed and implemented. These standards should include measures to protect against Denial of Service (DoS) attacks, malware infections, and other threats.

Fortunately, Artificial Intelligence (AI) can play a key role in mitigating these risks. With so many interconnected devices, it can be difficult to identify and isolate malicious traffic. AI can help by identifying patterns in data that would otherwise be undetectable to humans.

6G technology is still in the early developmental stages, but security experts are already voicing concerns about the potential challenges that could arise with this next generation of mobile connectivity. Experts are already working on a roadmap for 6G deployment, and they are confident that these and other challenges can be overcome.

5G Cyber Risks and Mitigation

Sabhyata Soni

CRC Press
Taylor & Francis Group
Boca Raton London New York

CRC Press is an imprint of the
Taylor & Francis Group, an **informa** business

First edition published 2023
by CRC Press
6000 Broken Sound Parkway NW, Suite 300, Boca Raton, FL 33487-2742

and by CRC Press
4 Park Square, Milton Park, Abingdon, Oxon, OX14 4RN

Library of Congress Cataloging-in-Publication Data

Names: Soni, Sabhyata, author.
Title: 5G cyber risks and mitigation / Sabhyata Soni.
Other titles: Five G cyber risks and mitigation
Description: First edition. | Boca Raton, FL : CRC Press, 2023. |
Includes bibliographical references and index.
Identifiers: LCCN 2022044726 (print) | LCCN 2022044727 (ebook) |
ISBN 9781032206127 (hbk) | ISBN 9781032206134 (pbk) |
ISBN 9781003264408 (ebk)
Subjects: LCSH: 5G mobile communication systems--Security measures.
Classification: LCC TK5103.25 .S66 2023 (print) | LCC TK5103.25 (ebook) |
DDC 005.8--dc23/eng/20221129
LC record available at https://lccn.loc.gov/2022044726
LC ebook record available at https://lccn.loc.gov/2022044727

ISBN: 978-1-032-20612-7 (hbk)
ISBN: 978-1-032-20613-4 (pbk)
ISBN: 978-1-003-26440-8 (ebk)

DOI: 10.1201/9781003264408

Typeset in Minion
by KnowledgeWorks Global Ltd.

Contents

About the Author, vii

About the Author

Dr. Sabhyata Soni is a PhD (Electronics & Communications Engineering) from Thapar University Patiala. She has more than 20 years of experience teaching in the topmost engineering colleges in India. Her research papers are published in many reputed and highly ranked international journals. Currently, she is teaching as a senior faculty at the University Institute of Engineering and Technology (UIET), affiliated with one of India's oldest universities, the Panjab University (PU), Chandigarh.

CHAPTER 1

Overview of 5G Network, Architecture, and Uses

5G is the next significant advancement in cellular networks from earlier Long-Term Evolution (LTE), Universal Mobile Telecommunication System (UMTS), and Global System for mobile Communication (GSM) technologies. Unlike 4G, also known as LTE, and 2G, also known as GSM, 5G is simply referred to as 5G. 5G enables very high-speed throughput, ultra-low latency, and a more significant number of connected devices. With these increased capabilities, 5G can support various applications, including Augmented Reality (AR), Virtual Reality (VR), IoT (Internet of Things), autonomous driving, 4K streaming, and more. With its high speed, low latency, and large capacity, 5G Ultra Wideband might make services such as cloud-connected traffic control, drone delivery, and other applications live up to their potential. The possibilities are nearly endless, ranging from emergency response to worldwide payments to next-generation gaming and entertainment.

THE TECHNICAL BASIS OF 5G

5G is based on OFDM (orthogonal frequency-division multiplexing), a technique for reducing interference by modulating a digital signal across many channels. 5G employs a 5G New Radio (NR) air interface in conjunction with OFDM principles. Broader bandwidth technologies like

DOI: 10.1201/9781003264408-1

sub-6 GHz and Millimeter-Wave (mmWave) are also used in 5G. The following are the technical terms used in 5G:

Millimeter-Wave: This term refers to spectrum bands with frequencies greater than 24 GHz. The abundant spectrum accessible at these high frequencies can enable extremely high data speeds and capacity, completely changing the mobile experience.

Massive multiple inputs and multiple outputs (MIMO): Uses huge antenna arrays at base stations to simultaneously service many autonomous terminals. Smart processing at the array takes advantage of the terminals' diverse and distinct propagation fingerprints.

Beamforming: Beamforming is a sort of RF management. Here, an access point sends out the same signal via numerous antennas. It ensures a smooth data delivery path to a user while minimizing interference to other users. Beamforming, in conjunction with massive MIMO, enhances spectrum efficiency and capacity.

Full duplex: This allows us to send and receive data simultaneously on the same channel. Advantages are more spectrum efficiency, symmetric fading characteristics, better filtering, new relay solutions, and improved interference coordination.

Small cell: Small cells are miniature base stations with little power. They use cellular or Wi-Fi technology to operate in licensed or unlicensed airwaves. 5G can achieve 1000× throughput with the help of small cells.

BASIC RADIO ACCESS NETWORK (RAN) ARCHITECTURE

Before moving on to 5G RAN, it is necessary to know RAN basics. A RAN is a critical component of a wireless telecommunications system that uses a radio link to connect individual devices to other network elements. Over a fiber or wireless backhaul connection, the RAN connects user equipment (UE) such as a cellphone, computer, or any remotely operated equipment. This link connects to the core network, which handles subscriber data, location, etc.

The radio part of the cellular network is the RAN, also known as the access network. A cellular network consists of cells, which are small land areas. A cell is served by at least one radio transceiver, though cell locations are usually operated by three transceivers.

From the 1G to the (5G of cellular networking, RANs have evolved. The 3rd Generation Partnership Project brought LTE RAN with the advent of

4G technology in the 2000s. The radio access network and core network got altered dramatically. System connectivity was based on the Internet Protocol (IP) for the first time with 4G, replacing earlier circuit-based networks.

Improvements in centralized RAN, also known as cloud RAN (C-RAN), and multiple antenna arrays, such as MIMO, are now available with LTE Advanced and 5G.

Components of a RAN

Depending on their capability, base stations and antennas that cover a specified region are RAN components. Silicon chips provide RAN functionality in both the core network and the user devices.

A RAN consists of three key components:

1. Antennas are devices that transform electrical signals into radio waves.

2. Radios convert digital data into wireless signals and guarantee that broadcasts are made at appropriate frequency ranges and power levels.

3. Wireless communication is made feasible by baseband units (BBUs), which provide a set of signal-processing operations. Traditional baseband enables wireless communication by combining bespoke electronics with several lines of code, which is usually done over licensed radio bandwidth. BBU processing detects mistakes, protects the wireless transmission, and assures efficient use of wireless resources.

In 5G, What Exactly Is RAN?

For 5G cellular technology, the 5G NR standard is the most recent radio interface and radio access technology. The interface supports multiple frequency bands, including sub-6 GHz and mmWave bands such as 24 GHz, 28 GHz, and higher bands. Compared to sub-6 GHz services, the mmWave bands offer download speeds of 1+ gigabits per second but have shorter ranges.

THE WORKING OF RAN

We'll now draw out the job of each base station in the RAN. Remember that this is similar to describing the Internet by explaining how a router works— it's a reasonably good place to start. Still, it doesn't do the entire story right. The whole story can be mentioned in six parts as mentioned below:

1. Upon power-up or handover when the UE is active, each base station establishes the wireless channel for a subscriber's UE (Figure 1.1). When the UE sits idle for a set amount of time, this channel is freed.

This wireless channel is described as a bearer service provider in 3GPP terminology. The term "carrier" has traditionally been used in telecommunications to refer to a data channel rather than a signaling channel (including early wireline systems like ISDN)

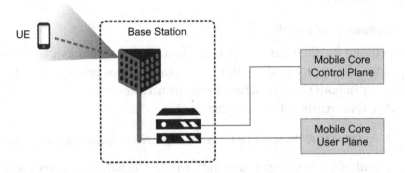

FIGURE 1.1 Base station detects (and connects to) active UEs.

2. Each base station provides "3GPP Control Plane" connectivity between UE and the mapped Mobile Core Control Plane component (Figure 1.2) and forwards signaling traffic between the two. UE authentication, registration, and mobility tracking are all enabled via this signaling stream.

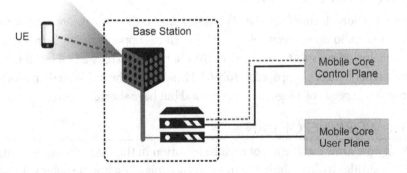

FIGURE 1.2 Base station establishes control plane connectivity between each UE and Mobile core.

3. The base station connects and maintains one or more tunnels between each active UE's corresponding Mobile Core User Plane component (Figure 1.3).

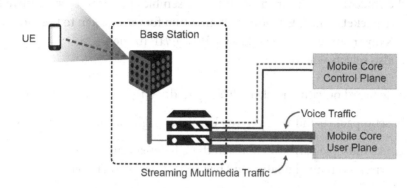

FIGURE 1.3 Base station establishes one or more tunnels between each UE and Mobile core's user plane.

4. Between the Mobile Core and the UE, the base station sends both control and user plane packets. Stream Control Transport Protocol (SCTP)/IP and General Packet Radio Service [GPRS] Tunneling Protocol (GTP)/UDP/IP are used to tunnel these packets. SCTP is a secure alternative to TCP designed to transfer telephony signaling (control) information. General Packet Radio Service [GPRS] Tunneling Protocol (GTP) is a 3GPP-specific tunneling protocol that runs over UDP.

It's worth noting that connectivity between the RAN and the Mobile Core is based on IP. Introducing this feature was one of the most significant differences between 3G and 4G. Before 4G, the cellular network's internals was circuit-based, which is understandable given the network's roots as a speech network.

Each base station uses direct station-to-station communications to coordinate UE handovers with surrounding base stations. These connections are used to transfer both control plane (SCTP over IP) and user plane (GTP over UDP/IP) packets, just like the station-to-core connectivity illustrated in the preceding figure.

5. Base stations coordinate multi-point wireless transmission to a UE from many base stations, which could or could not be part of a UE handover between base stations.

The most crucial point is that the base station can function as a customized forwarder. It fragments outgoing IP packets into physical layer segments. It schedules them for transmitting over the available radio spectrum in the Internet-to-UE direction. It also assembles physical layer segments into IP packets. The UE-to-Internet direction forwards them to the Mobile Core's upstream user plane (via a GTP/UDP/IP tunnel).

It also selects whether to:

a. forward outgoing packets directly to the UE

b. send packets indirectly to the UE via a neighboring base station, or

c. use multiple pathways to reach the UE depending on observations of wireless channel quality and per-subscriber regulations.

In the third situation, the physical payloads could be distributed across numerous base stations or different carrier frequencies of a single base station (including Wi-Fi).

ABOUT THE MOBILE CORE

A mobile core network is a critical component of a more extensive mobile network that allows users to access the services to which they are entitled. It is in charge of essential functions such as subscriber profile information, location, service authentication, and switching activities.

The Mobile Core's main job is to offer external packet data network (i.e., Internet) connectivity to mobile subscribers while guaranteeing that they are authorized and that the observed service quality meets their subscription SLAs (Service level agreements). The Mobile Core must manage all subscribers' mobility by keeping track of their last whereabouts at the granularity of the serving base station, which is a significant component. The Mobile Core's architecture is complicated because it keeps track of individual subscribers, which the Internet's core does not. This is especially true given that those subscribers move around.

Note: While the overall functionality of 4G to 5G remains roughly the same, the functionality is virtualized and its integration into individual components varies.

The cloud's move toward a microservice-based (cloud-native) architecture greatly influences the 5G Mobile Core. This move to cloud-native is more significant than it appears at first glance because it allows for greater flexibility and specialization. The 5G Mobile Core could evolve to serve more than simply voice and broadband connectivity, such as large IoT, which has a fundamentally different latency need and usage pattern (i.e., many more devices are connected intermittently).

5G MOBILE CORE

While the 3GPP specification specifies this disaggregation level, it only prescribes a set of functional blocks and no implementation. The 5G Mobile Core, which 3GPP calls the NG-Core, adopts a microservice-like architecture. The collection of technical considerations that develop a microservice-based system differs substantially from a set of functional blocks. However, considering the components in Figure 1.4 as a collection of microservices is a suitable working paradigm.

Note: "microservice-like" means, while the 3GPP specification specifies this level of disaggregation, it is more of a set of functional blocks than a complete implementation.

The collection of functional blocks is divided into three groups, as shown below.

The first group operates in the Control Plane (CP), while the second operates in the Evolved Packet Core (EPC).

The first group operates on the CP and has an EPC counterpart:

EPC Counterparts

AMF (*Access and Mobility Management Function*): Connection and reachability management, mobility management, access authentication and authorization, and location services are all responsibilities of this position and manage the EPC's MME's mobility-related elements.

Session Management Function (*SMF*): Controls all aspects of each UE session, including IP address assignment, UP function selection, Quality of Service (QoS) control, and UP routing control. Roughly corresponds to a portion of the EPC's MME and the EPC's PGW's control-related features.

Policy Control Function (*PCF*): Manages the rules of policy that are then enforced by other CP services. Approximately correlates to the PCRF of the EPC.

Unified Data Management (*UDM*): Manages the identity of users, including authentication credentials creation. Part of the functionality of the EPC's HSS is included.

Authentication Server Function (AUSF): An authentication server in essence. Part of the functionality of the EPC's HSS is included.

The second group similarly operates on the CP but does not have an EPC counterpart:

Structured Data Storage Network Function (SDSF): An "assistant" service for storing structured data. In a microservices-based system, it might be implemented by a "SQL Database."

Unstructured Data Storage Network Function (UDSF): An unstructured data storage "support" service. In a microservices-based system, it could be implemented as a "Key/Value Store."

Network Exposure Function (NEF): A way to make certain capabilities available to third-party services, such as data translation between internal and external representations. In a microservices-based system, it might be implemented by an "application programming interface (API) Server."

NF Repository Function (NRF): A way to determine what services are available. In a microservices-based system, it may be implemented by a "Discovery Service."

Network Slicing Selector Function (NSSF): A function that allows you to choose a Network Slice to serve a specific UE. Network slices are a method of partitioning network resources to diversify services given to various users.

The third group includes only one component that runs in the User Plane (UP):

User Plane Function (UPF): This function forwards traffic between the RAN and the Internet and corresponds to the S/PGW combination in EPC. It is also in charge of policy enforcement, lawful intercept, traffic consumption reporting, and QoS policing, in addition to packet forwarding.

The first and third groups are best understood as a simple restructuring of 4G's EPC, while the second group is 3GPP's method of pointing to a cloud-native solution as the planned end-state for the Mobile Core, despite the superfluous inclusion of new terminology. It's worth noting that creating separate storage services allows all other services to be stateless and hence more easily scaled. Also, instead of showing a complete set of pairwise connections, Figure 1.4 depicts a message bus connecting all components, which is a popular idea in microservice-based systems. It also indicates that the implementation technique is well-understood.

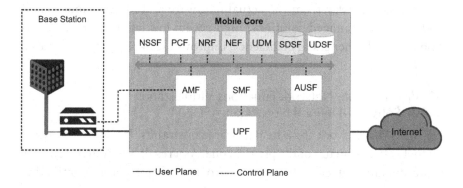

FIGURE 1.4   Mobile Core (NG-Core).

5G VIRTUALIZATION

As 5G is not the same as the previous Generations, this technology does not simply require a network upgrade; it necessitates the creation of an entirely new network.

Initially, 5G will need to integrate existing 4G operations, particularly LTE. Still, operators will need to reinvent their network architecture, processes, and services to scale their networks quickly and access 5G's full potential.

Virtualization of 5G networks is critical and unavoidable.

5G will be unable to achieve its connectivity requirements without virtualization and as such the network will be unable to keep up with the rapid technological advances occurring in external/peripheral fields and further the investments made by telcos will not yield a profit.

5G virtualization replaces hardware-based network services with software-based network functions to optimize the network while boosting the infrastructure's efficiency, flexibility, and capacity.

In cloud computing and enterprise environments, virtualization of services and network operations has increased in popularity over the last decade to provide high availability through self-deployment, self-scaling, and self-healing of services. This is especially true in 5G mobile broadband networks, which are virtualizing 5G core architecture, transport networks, and even radio access technologies to run on shared infrastructure.

This virtualization enables a flexible environment where services can be deployed across several geographic locations and dynamically instantiated to reduce system load.

Note: 5G will only be able to unleash its full range of latent benefits by moving away from the actual hardware and toward cloud-based and software-managed solutions: high speed, low latency, lower operational costs, higher energy efficiency, improved scalability, and increased agility.

5G Virtualization, Network Functions Virtualization (NFV), Software-Defined Networking (SDN)

NV frees the network from its hardware constraints. It allows it to run a virtual network on top of the physical one. As a result, the system becomes more dynamic. It can be managed from a central plane, eliminating humans' need to manually configure the hardware.

The hardware resources are divided into functions so that managing software will be possible with 5G NV, which is NFV. NFV aims to optimize network services in network management directly. The accompanying network management technique, SDN, creates a centralized network view by decoupling the control and forwarding planes.

Network resources can be customized and assigned to serve the needs of specific consumers or service categories with NFV without the need for physical changes or dedicated infrastructures.

This reorganization will pave the path for much-touted 5G capabilities such as network slicing. This architecture allows numerous virtual networks to exist on top of the physical shared infrastructure.

Each network slice can be mapped to specific operations, clients, or use cases, resulting in improved service at each leg and a more efficient network overall. The ability of 5G to support and deliver value from the three ITU-specified generic services with dramatically different requirements will rely heavily on network slicing.

USE CASES

Although the list of 5G applications seems limitless, most cases fit into one of three categories (Table 1.1):

1. eMBB (Enhanced Mobile Broadband)

2. URLLC (Ultra-reliable and Low-latency Communications)

3. mMTC (Massive Machine Type Communications)

TABLE 1.1   5G Use Cases

5G Use Case	Example	Requirements
eMBB	UHD video, gaming	High capacity, video caching
mMTC or massive IoT (mIoT)	Large networks of sensors for smart cities, agriculture, etc.)	A large number of connections — over 1 million per km². Mostly immobile devices
URLLC or mission-critical IoT	Autonomous transportation, smart manufacturing, smart grid, public safety, etc.	Low latency (ITS 5ms, motion Control 1 ms), high reliability

Note: It would make no commercial sense to build various physical networks to service different sorts of use cases.

High-speed mobile apps, driverless autos, and mIoT will all develop from these three sectors. Notably, the network requirements including speed, latency, stability, and security for each use case will vary. Network slicing allows these needs to be met in a scattered yet coordinated and personalized manner.

Automation, service provisioning, and orchestration are used to personalize the network offering, which SDN and NFV support. However, decoupling hardware and software does more than just improving the network efficacy and efficiency.

It offers itself a more democratic approach to wireless innovation, offering better services, better network economics, and quicker time to market for new network providers.

CONSIDERATIONS FOR NETWORK PROGRAMMABILITY IN 5G NETWORK

Programmability is a notion that entails network softwarization and virtualization through the use of SDN/NFV architecture. 5G programmability necessitates a deliberate separation and abstraction of NFs. 5G programmability enables the rapid, flexible, and dynamic deployment of new network and management services in all network segments as groups of virtual machines (VMs) (control and management plane). By employing an open API and Software Development Kits, 5G programmability will promote the building of 5G ecosystems that might benefit various control and management planes intuitively network-wide (SDK).

ABOUT NETWORK SLICING

The Next Generation Mobile Network (NGMN) was the first to coin and implement 5G network slicing. According to the NGMN, a network slice is an end-to-end (E2E) logical network/cloud that is mutually isolated, runs on a similar underlying (physical or virtual) infrastructure, and has autonomous control and administration that may be constructed on demand. A network slice can be made up of cross-domain components from multiple domains under the same or distinct administrations and components for access, transport, core, and edge networks. Network slices are self-contained, mutually separated, controllable, and programmable to support multi-service and multi-tenancy.

The 3rd Generation Partnership Project (3GPPP) defines network slicing as a "technology that enables the operator to design customized networks to deliver efficient solutions for different market scenarios with varying requirements (e.g., in terms of functionality, performance, and isolation)." From a business perspective, a slice is a collection of all necessary network resources, functions, and assets required to complete a certain business case or service, including Operation Support Systems (OSS), Business Support Systems (BSS), and DevOps processes. So, there are two types of slices: (a) internal slices, which are partitions used for the provider's internal services, over which the provider has complete control and management and (b) external slices, which are partitions hosting customer services and appear to the customer as dedicated networks/clouds/data-centers. Network slicing can deliver radio, cloud, and networking resources to application providers and vertical segments that do not have physical network infrastructure. As a result, service differentiation is enabled by tailoring network operations to match customers' needs based on the type of service.

5G NETWORK SLICING

One of the key differences between 5G and previous technologies is that the network architecture will no longer consist of homogeneous parts. Instead, the architecture is built on the network slicing concept, which employs NV and softwarization of various network elements. Network slicing allows numerous virtual networks to be built on top of shared physical infrastructure.

Network slicing is yet another widely anticipated feature of 5G. "According to Inskeep, this allows for dedicated resource blocks on the RAN. According to Inskeep, 5G might maintain particular performance thresholds in each slice of the cellular network, an attribute that was previously only available on the wired side."

"These automobiles and drones collect a lot of data, which must be analyzed rapidly so that they can be useful. 5G and edge computing's end-to-end speed and decreased latency, along with network slicing to provide data its own 'lane,' will let data travel back to the device in substantially less time."

With 5G, one can theoretically build a single platform capable of tackling IoT use cases that were previously covered by combining wired and wireless networks. With 5G, one can answer most of those use cases with wireless from a single platform.

This should be persuasive for firms wanting to increase the value of IoT from a management and operating expense perspective, according to Filkins.

A "network slice" is defined by the 5G Infrastructure Public-Private Partnership (PPP)16 as a bundled collection of network services, applications, and infrastructure. A network slice might be assigned to specific applications, services, use cases, or business models to satisfy their needs. It can be devoted to a single tenant (a service provider) who uses it to deliver a specific telecommunication service (e.g., eMBB). Decoupling the virtualized and physical infrastructure enables slices to scale efficiently.

Network slices cover the entire protocol stack, from the underlying (virtualized) hardware resources up to the network services and applications that run on top of them. A network slice, comprising OSS and BSS, provides a combination of all relevant network resources, network, and service functionalities required to fulfill a specific business case or service.

To put it another way, network slicing enables operators to "slice" a single physical network into many virtual E2E networks across the device, access, transport, and core networks. Each slice is logically isolated, with any faults or security issues confined inside it. Each slice has specialized resources, such as network bandwidth or QoS, that are customized for various types of services with multiple features and requirements.

In terms of business, network slicing enables a mobile operator to develop customized virtual networks for specific clients and use cases. Different parts of 5G technology will help specific applications, such as mobile broadband, machine-to-machine communications (e.g., manufacturing or logistics), or smart automobiles. Higher speeds, lower latency, and access to edge computing resources may be required by everyone. A 5G operator can provide personalized solutions to specific businesses by designing different slices that prioritize particular resources. Some sources claim that this will revolutionize industries such as marketing, augmented reality, and mobile gaming, while others are more skeptical, citing patchy network coverage and the limited reach of benefits other than higher speed.

Slicing can also improve service continuity by allowing a host network to create an optimized virtual network that replicates the one offered by a roaming device's home network. It can also do it by enabling a host network to create an optimized virtual network that replicates the one offered by a roaming device's home network.

5G NETWORK SLICING ENABLING TECHNOLOGIES

SDN is a method of bringing intelligence and flexibility to 5G networks, allowing for more fine-grained and network-wide orchestration and control of applications and services.

The Open Networking Forum defines SDN as the physical separation of the network control plane from the forwarding plane, with each control plane controlling multiple devices. The main functions of SDN are:

- Separation of the control and data planes

- Providing centralized control and network programmability

- Usability in data center servers and switches

- Usability of OpenFlow protocol for communication

- Usability in networking

- Standardization by the Open Networking Forum

Benefits of SDN

- Provides complete network control

- Scales without affecting performance

- Network virtualization

- Controlling data flow in the network to reduce network congestion

- Provides customized services to its customers

- Provides a secure communication channel

- Convenient integration of new IoT devices

- Each customer has an isolated view of the network

- Device configuration and troubleshooting are done from a single point

Network Function Virtualization (NFV)

Virtualizing network services (Figure 1.5) to keep them running on proprietary, specialized equipment is known as NFV. Routing, load balancing, and firewalls are combined as VMs on physical devices using NFV and they are

- Used in service providers' networks

- Used in firewalls, gateways, and content delivery networks

- Standardized by the European Telecommunications Standards Institute NFV Group

Benefits of NFV

- Creates multiple networks on the same platform

- Isolation between networks or malfunction in one network will not affect the production

- Packet handling

- Quality of Service

- Transport efficiency

- Resource optimization

- It cuts down CAPEX and OPEX

- Scalable

- Load balancing

- High performance

Role of SDN in 5G

SDN is a smart network architecture that aims to reduce hardware constraints. SDN's goal is to abstract lower-level functions and shifts them to a standardized control plane that regulates network behavior through APIs. Network administrators can supply services via the network, notwithstanding the associated hardware components using a software-based, centralized control plane.

With the available spectrum, 5G will push the envelope of what is possible. That's where SDN comes into play in the 5G world. SDN can be used to create an overall framework for 5G to work across a control plane. Data passing throughout the 5G network can deliver better data flows. Furthermore, SDN architecture can save network bandwidth while increasing latency. Finally, because SDN can be utilized in 5G networks, it offers a centralized control plane to monitor and automate network redundancy, avoiding major outages by identifying appropriate data flows in real time.

- The network will be able to manage the data flow in the network thanks to SDN.

- SDN will improve the server-to-server traffic.

- SDN will reduce latency and improve network congestion.

- Through a secure route, SDN will enable faster and easier integration of new devices.

- SDN will make it possible to configure and troubleshoot devices from a single location.

Role of NFV in 5G

The main idea underlying NFV is to disconnect software from hardware, even though it is still a young technology. Service providers can use NFV to install various network functionalities on VMs, such as firewalls and encryption. When a customer requests a new network function, service providers can immediately create a VM for that function. Network administrators do not need to invest in high-priced, proprietary hardware to set up a service chain of network-connected devices when using this technology. These network functions, unlike proprietary hardware, can be installed in weeks rather than in months.

In the case of 5G, NFV will aid in the virtualization of many network appliances. NFV will enable 5G network slicing, which will allow many virtual networks to run on top of a single physical infrastructure. Furthermore, 5G NFV will allow the division of a physical network into many virtual networks capable of supporting multiple RANs. By optimizing resource provisioning of VNFs for price and energy, scaling VNFs, and ensuring VNFs continuously operate effectively, NFV can help solve barriers to 5G.

- NFV allows for hardware independence and the creation of many networks on a single piece of hardware.

- NFV allows for load balancing and dynamic resource allocation based on demand.

- Once virtualized, NFV will use the physical network for packet filtering.

- NFV allows for quick scaling without sacrificing the performance.

- NFV allows networks to be isolated from one another. The production will not be harmed if one of the components fails.

5G NFV: KEY TAKEAWAYS

1. NFV is the process of converting hardware-based networking functions into software that can run on standard hardware.

2. NFV is employed in 5G networks since the industry and standards organizations have determined that 5G must-have software advantages that match their expectations.

3. Network slicing in 5G networks is enabled by NFV, which allows a virtual network to be built on top of physical infrastructure and subsequently partitioned for different applications and uses based on demand.

4. Scalability, robustness, and fault tolerance are advantages of using NFV technology in a distributed cloud.

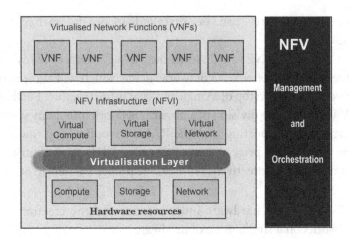

FIGURE 1.5   High-level NFV framework.

NFV VS. SDN

SDN is a network architecture that injects automation and programmability into the network by decoupling network control and forwarding functions. We can say that NFV refers to the virtualization of network components, while SDN is a network architecture that brings automation and programmability to the network by decoupling network control and forwarding functions.

To put it another way, NFV virtualizes network infrastructure. In contrast, SDN centralizes network management. SDN and NFV work together to build a network that is constructed, controlled, and managed entirely by software.

An SDN typically includes an SDN controller for northbound and southbound APIs. The controller allows network administrators to see the network and regulate the underlying infrastructure's behavior and policies. Southbound APIs gather network information from that infrastructure and send them back to the controller, which is required to keep the network working smoothly. Northbound APIs are used by applications and services to express their resource requirements to the controller.

5G and Cloud Computing

The advent of 5G will spur cloud architectures based on virtualization and microservices to accelerate their growth. Providers of cloud-based software virtualization technology and knowledge and technology suppliers focused on service automation, security, and cloud-native designs would gain a lot from this.

Mobile Cloud Computing (MCC), Multi-access Edge Computing (MEC), Cloud Radio Access Network (C-RAN), and other innovative designs and technologies revolutionize the cloud computing environment. It is predicted that an "all-cloud 5G architecture" will be developed, allowing for an all-cloud digital transformation of networks, OSS/BSS, and services.

Making 5G Work: Cloudification and 5G

The link between 5G technology and cloud computing has two significant features. First and foremost, cloud computing must evolve to meet the demands of 5G. The expanding importance of edge, mobile edge, and fog computing in cloud computing reflect this. The second factor is that 5G technologies are being "cloudified" as a result of network "softwarization", NFV, SDN, and other technologies. As a result, both technologies impact each other's progress. The adoption of 5G raises questions about the convergence of computing, cloud, and IoT, ushering in a new era of hyper-connectivity. In this paper, I examine both aspects of the evolution of 5G and cloud technologies, focusing on interdependencies.

Will 5G Technologies Destroy Cloud Computing?

According to some industry analysts, 5G could be a cloud computing killer, rendering the cloud as obsolete. There's even talk of a "post-cloud world." For example, Jon Markman stated in a Forbes article that ultra-fast wireless networks would eventually supplant the cloud as a computing platform. Hundreds of billions of autonomous smart devices would populate a post-cloud world. These devices will process the data collected at the network edge in real time.

5G and Cloud Computing Work in Harmony

Because 5G devices can only process local data, cloud computing with adaptable computer resource pools is essential for 5G technology-based services that manage compute- and data-intensive applications like Big Data analysis and Artificial intelligence (AI). Cloud computing has evolved to include MCC and a multi-access edge cloud computing environment to fulfill 5G expectations. As a result, the designs and management frameworks for next-generation cloud technologies will need to change dramatically. In addition, the cloud is regarded as the cornerstone of the C-RAN. Cloud computing and 5G networks are both undergoing significant changes. Cloud platforms are altered to fulfill 5G requirements and reshape the telecoms' architectural environment by providing flexibility, lower total costs, and high service availability. The capabilities of the cloud will be extended to the network edge, forming an edge cloud similar to MEC. This allows network services such as C-RAN to run at the network's edge. The future network architecture will combine edge and core clouds, resulting in distributed cloud computing.

HOW WILL THE 5G NETWORK BENEFIT CLOUD COMPUTING?

Let us check how the 5G network will benefit cloud computing.

It Will Facilitate Faster Streaming of Data and Analytics

Any firm or entity that uses a 5G network will benefit from faster data transfers and analytics than they have ever experienced before. 5G networks are well-suited to allowing real-time streaming from one location to another. Businesses that deal with time-sensitive and big data technology, such as metrological and space agencies, can gain significantly from this. Storage and streaming will take place in real time, which is the pinnacle of production.

It Will Simplify Internet-Based Industrial Operations

The streamlined working style of the 5G network will assist any industry that uses the cloud to run its operations. The network's high speeds and dependability will benefit a supply chain management sector and a cloud computing security company can use vast amounts of sensitive data in real time. As the business's operations can be quantifiably planned, the burden is bound to become more manageable. From now on, order tracking, inventory control, and delivery tracking, among other things, will all be easily accessible from a single platform.

Because of improved connectivity, work can be done from any location. In the past, you had to report to an office to complete a task. In today's world, you may do the same work and much more from the comfort of your own home. You can do it conveniently from anywhere as long as you have a stable connection. This experience is enhanced by cloud computing technology coupled with a 5G network. This is because the sensors on the connected remote devices use this high-speed, low-latency network to get the required information on the fly. Thanks to this technology, companies and individuals can track their cargo or cars as they move. It will also facilitate the development of better security automation systems for homes and offices. Every industry saves a lot of money and effort.

It Will Considerably Boost AI Technologies' Capabilities

AI technology is still in its infancy. The advancements made since its discovery have been revolutionary, yet they have yet to reach their pinnacle. Given that the ability of a computer to impersonate a human is the main driver of an AI system, processing speed is also critical. Because of this ability, robotics in industry and healthcare, for example, would benefit significantly from the 5G network. A surgeon can now use a robot to operate on a patient in another area. The process will improve further with a 5G network. Virtual and augmented reality technologies will also benefit for the same reason.

It Will Encourage the Development of More Effective Security Solutions

Perpetrators and hackers are the drawbacks that come with the ever-evolving webspace. Such actors undermine many collaborations by penetrating firewalls and gaining access to companies' and governments' essential and sensitive information. However, administrators will be able to spot and avert such cyber-attacks with 5G connectivity. Hybrid cloud computing is comprehensively safe, with the added benefit of smoother cloud-to-cloud transfers.

CHAPTER 2

5G Use Cases and Application

CATEGORIZING THE APPLICATIONS FOR 5G

5G will be transformative and will enable many new applications that are not viable today, especially in urban areas and cities. 5G use cases (Figure 2.1) are not limited to a specific area: consumers, businesses,

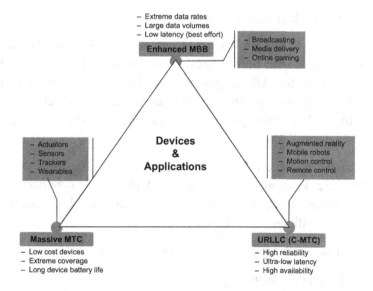

FIGURE 2.1   eMBB, mMTC, and URLLC are different use cases supported in 5G NR technology.

DOI: 10.1201/9781003264408-2   23 industries, and cities will benefit from one or more dimensions of the 5G triangle as shown above.

5G Use Cases

The following use cases are supported by 5G networks, which are a cornerstone in developing infrastructure to deploy new business models:

1. Ultra-reliable and low-latency communications (URLLC): URLLC is required for applications such as driverless vehicles, smart grids, industrial automation, remote patient monitoring, and telemedicine. The 5G URLLC use case supports the following features.

   - It provides ultra-responsive connections.

   - It offers lesser than 1 ms air interface latency

   - It offers 5 ms end-to-end latency between UE (i.e., mobile) and 5G (Evolved NodeB) eNB (i.e., base station).

   - It is ultra-reliable, and availability is 99.9999% of the time.

   - It offers low to medium data rates (about 50 kbps to 10 Mbps).

   - It offers high-speed mobility.

   URLLC or Mission-Critical Control use cases require extremely low latency with high reliability, availability, and security. Such cases can be AR (augmented reality)-enabled maintenance, AR-supported tourism, autonomous vehicles or vehicles to everything (V2X) cases, product line automation, remote operations, and telemedicine.

2. Massive machine-type communications (mMTC): The potential to handle at least one million Internet of Things (IoT) connections per square kilometer, with broad coverage within buildings. 5G mMTC use case will have the following features.

   - It supports a high density of devices (about 2×105 in $106/Km^2$)

   - It supports long-range.

   - It supports a low data rate (about 1 to 100 Kbps).

   - It leverages the benefits of the ultra-low-cost of M2M.

- It offers ten years of battery life.

- It provides asynchronous access

mMTC use cases are cases where low-power, low-data devices are connected at scale, for example in smart cities, as autonomous vehicle assistants, remote-controlled robot controls, or port/airport operations. Narrowband IoT improvement is needed to take full advantage of mMTC, such as language support, location support, device mobility, and over-the-air firmware updates. Efficient uplink transmission with multiple access in the form of a WAN-managed multi-hop mesh architecture is also required.

3. Enhanced mobile broadband (eMBB): It offers theoretical peak download speeds of up to 20 Gb/s and a consistent data rate of 100 Mb/s in urban areas. This will make it easier to handle growing video consumption and new virtual and AR services. The following features are supported by the 5G eMBB use case.

- The peak data rate is 10 to 20 Gbps.

- An Internet speed of 100 Mbps can be achieved when required.

- The traffic will be 10,000 times more.

- It supports macro and small cells.

- It supports high mobility of approx. 500 Kmph.

- It helps in saving network energy by 100 times.

eMBB can be subdivided into numerous sub-use-cases, for example, Fixed Wireless Access (FWA), video downstream/video upstream, etc. eMBB will be the focus of the first 5G deployments. eMBB use cases are data-intensive use cases that require more bandwidth. These cases include cloud and UHD, 8K video streaming, immersive gaming (including AR and VR [virtual reality]), video analytics, immersive event experience, and telemedicine.

Note: FWA can provide fiber-like speeds to households and businesses by using new frequency bands, mMIMO, and 3D beamforming technologies.

URLLC—High Availability, Low Latency Use Cases for 5G

Once 5G is fully deployed, URLLC will be one of the most significant game-changers. New apps that demand response in fractions of a second will be seen here.

Autonomous Vehicles

One of the most anticipated 5G applications is autonomous vehicles. Vehicle technology is fast improving to assist the future of autonomous vehicles. Onboard computer systems are improving to the point where they can now match the processing capacity of data centers.

Today, we hear about self-driving cars. Many people wonder what are the obstacles to making this futuristic technology a reality. For a completely autonomous vehicle future to become a reality, various advancements in-vehicle technology, network speed, data throughput, and machine learning (ML) must all come together.

Because of significantly reduced latency, 5G networks will be a massive enabler for autonomous vehicles because vehicles will be able to react 10–100 times faster than on existing cellular networks.

The ultimate goal is to create a V2X communication network. Vehicles will be able to react almost immediately to elements and changes in their environment. A vehicle must be able to communicate and receive data in milliseconds in order to stop or change directions in response to road signs, risks, and pedestrians crossing the street.

Differences Between 4G and 5G Latency

Let us understand it with an example. Suppose a car traveling at 30 miles per hour needs to receive a signal to avoid colliding with an object. An automobile would go roughly 4 feet or 1.2 meters at current 4G latency of around 100 milliseconds. The car would have only driven 5 inches (12 cm) due to approximately a 5G delay of 10 milliseconds. The difference is huge and it might mean the difference between life and death.

5G IoT Infrastructure and Traffic Management in a Smart City

Many cities worldwide are presently implementing intelligent transportation systems (ITS) to enable connected car technologies. These systems are relatively straightforward to build, as they use existing communications networks to provide smart traffic management and the routing of emergency vehicles. Connected vehicle technology will enable bidirectional vehicle-to-vehicle (V2V) and vehicle-to-infrastructure (V2X) connections to increase safety across transportation networks. In smart cities, sensors are being deployed in every junction for motion detection and trigger connected and autonomous vehicles react as needed.

Note: The communications backbone for connected car technology might be installed now, long before 5G is fully deployed, dramatically enhancing pedestrian and vehicle safety.

5G IoT Applications in Industrial Automation

The key advantages of 5G in the industrial automation sectors include wireless flexibility, reduced pricing, and the feasibility of applications that are not thinkable with current wireless technology.

One would have probably seen demonstrations of synchronized robotics in manufacturing and supply chain applications. Wi-Fi lacks the range, mobility, and service quality required for industrial control, and cellular technology presently has too much latency for many applications. With 5G, industrial automation applications can become totally wireless, making smart factories more efficient. Industry 4.0 integrates the IoT and related services in industrial manufacturing and enables unified vertical and horizontal integration down the entire value chain and across all layers of the automation pyramid. Industry 4.0 will be supported by the powerful and pervasive connection between machines, people, and objects, which is a crucial component of the initiative.

Humans and robots will both be able to interact and work together in Industry 4.0. For example, a machine will carry heavy parts, and a person will be able to attach/assemble them. The robot must be in continual connection with the factory and its surroundings for this to happen. It must be mobile, with a full spectrum of physical motion and ambient sensors. These advancements will allow for symbiotic human-machine relationships in which each performs its optimal job.

Augmented Reality (AR) and Virtual Reality (VR)

Because of 5G's reduced latency, AR and VR applications will be far more engaging and dynamic. For example, a technician wearing 5G AR goggles could see an overlay of a machine that alongside identifying parts, also provide maintenance instructions, or show elements of the machine that are not safe to handle in industrial settings. There will be various chances for complex tasks to be supported by highly responsive industrial applications.

Enterprises can hold AR meetings in which two people appear to be in the same room. It will transform monotonous phone or 2D video conferences into more interactive 3D gatherings. Sporting events and experiences are projected to be among the most popular 5G consumer applications. When you react fast to stimuli, such as in a sports training application, you need to have the least amount of latency possible. For example, if two persons wearing 4G Long Term Evolution (LTE) goggles were attempting to kick a soccer ball back and forth, timing their response would be tricky since, by the time their brain receives the message that the ball has arrived, it will be too late. The decreased latency of 5G eyewear allows the receiver to see the ball and kick it back before it passes.

In sporting arenas, we will see increasingly immersive AR experiences. Virtual players will greet you and cheer you up as you go in if you have a 5G phone with AR. You'll also see larger-than-life replays and player stats during the game.

People are expected to see more hologram entertainers and greeters in the entertainment industry. For example, we can resurrect Elvis Presley or Patsy Cline using holograms. You can also make your own AR dancing partner.

5G IoT Applications for Drones

Beyond consumer videography and photography, drones now offer various applications. Utilities are increasingly using drones to assess equipment. Logistics and retail companies are considering drone delivery of goods. The trend will continue, and 5G will allow us to test the limits of today's drones, especially in terms of range and interactivity.

Currently, drones are limited to line of sight and controller distance. If you can't see the drone or it's out of range, you won't be able to control it or see where it's going. With 5G, you can put on glasses and "see" beyond current borders with low latency and high-definition video. 5G will also extend the range of controllers beyond a few kilometers or miles. These developments will benefit from rescue operations, border security, surveillance, drone delivery, and other applications.

Massive IoT Use Cases for 5G

One of IoT's main challenges will be its rapid expansion. According to Statistica, the number of IoT-linked devices per individual on the earth will rise from 2 per person currently to 10 per person by 2025. The anticipated number of connected gadgets that require a data connection puts enormous strain on communications infrastructure, such as cellular towers. While 4G is now satisfying this need satisfactorily in regions with a high cell density, 5G will improve this further.

The IMT-2020 standard, connected with 5G, demands a minimum connection density of 1 million devices per square kilometer, according to the minimum requirements documents (roughly 0.38 square miles). In comparison, the 4G LPWA (Low Power Wide Area) standard enables 60,680 devices at the same coverage size—a far cry from the capabilities of 5G.

Applications Benefitting from Massive IoT—Wearables and Mobile

What would be the catalyst for such a surge in linked IoT devices?

The Massive IoT element of 5G will be a massive market for wearables, trackers, and sensors. Consider the day when all of the gadgets, appliances, and equipment you interact with daily and your phones, tablets, and laptops are directly connected over a cellular connection. In any given location, 5G will allow significantly more devices to operate seamlessly (without perceived delays, dropped signals, and so on).

High-Speed Use Cases for 5G

We'll see a variety of applications in the high-speed use cases that are currently limited by slow speed. Consumers and businesses will access ultra-fast Internet via FWA.

Increased bandwidth applications such as 4K and, in the coming, 8K streaming or 360-degree video will allow consumers to experience high-quality, immersive experiences at real-time speeds. As a spectator, you will have complete control over the angle you want. For example, you can look about in a video car race to see who is close to or behind you.

Businesses can keep more data in the cloud and access it, as if it were stored locally, through 5G's fast, low-latency network. Now, It is not essential to have expensive on-site servers. To render information locally, you don't need a fast laptop; instead, you render this in the cloud and then have it streamed to you. It will be comparable to local, but you can access these high-end apps using your phone.

5G will also revolutionize the way businesses think about connectivity. Today, your organization may have a fiber, DSL, or cable modem line for primary connectivity and cellular backup if your primary connection fails. However, cellular can become your primary connection with 5G's high bandwidth, stability, and low latency. You won't have to bother about building wiring or the fees associated with its installation. You get the device, plug it in, and it works with cellular.

TEN MOST EXCITING 5G USE CASES

Here are ten of the most exciting 5G use cases.

1. Increased Agriculture Productivity

   Smart farming is already in progress, with enhanced computer capabilities and with IoT enabling data collection, analytics, and decision-making to reduce costs, resource usage, and increase yields. According to industry studies, 5G can expand the geographic reach of smart agriculture while also lowering costs by introducing high-capacity connectivity to rural farming areas.

2. Improved Distance Learning

   The risks of present connectivity infrastructure were highlighted by pandemic-era constraints that led to distant learning. Students have no choice except to rely on shaky and unreliable networks. More towns will have access to 5G's high speeds, increased capacity, and excellent dependability as telecom companies expand their 5G networks.

   As a result, 5G should make distant educational experiences more accessible. More crucially, academic institutions can use 5G to create and distribute new and unique learning content, such as live event streaming.

3. More Intelligent Logistics

   The logistics industry, which includes transportation, has been extending its use of IoT to track shipments as they travel throughout the US, across international borders, and the world. Autonomous vehicles in warehouses and on the road are also progressing in the business.

   However, the sheer number of sensors required to move and analyze all of that data puts a strain on 4G and LTE networks. As a result, the industry's ability to use innovative technologies is limited. This cellular capacity limitation is removed with 5G, allowing the industry to extend its use of smart devices.

4. High-Tech Medical Care

   Another area that is utilizing 5G to expand and improve its operations in healthcare. 5G can enable crucial and life-and-death use cases common in the medical field. 5G can be used in various ways in healthcare, including analytics, patient monitoring, remote diagnostics, and robot-assisted surgery. The technology rivals fixed-line networks in terms of dependability and capacity while allowing mobility within buildings that wired networks cannot.

5. More Efficient Manufacturing Processes

5G has unlimited potential for the manufacturing industry, as it promises greater flexibility than wired networks while still meeting the industry's high-capacity, high-reliability, and low-latency requirements. 5G enables automated manufacturing operations to be reconfigured more quickly to meet changing market demands.

6. Mining, Oil, and Gas Industries Modernization

Mining, oil, and gas activities in remote and difficult places can benefit from 5G. Due to cost and logistics, many facilities cannot construct wired networks and cannot rely on 4G/LTE networks to handle the massive volume of mission-critical data. More businesses are moving to 5G because it can handle the large-scale Industrial Internet of Things (IIoT) infrastructure required to monitor working conditions and direct automated gear. Moreover, when combined with edge computing, 5G might help the oil and gas industry make the most of the massive amounts of data generated by machinery.

7. Retail That Is More Personalized and Efficient

By combining digital capabilities with in-person services, retailers want to engage customers better and provide more customized experiences. To do so, many people are turning to next-generation technology, including 5G connectivity. An example of this is AR to help customers visualize furniture in their own homes, sensors feeding data to analytics systems to better manage inventory so customers don't find empty shelves, and personalized advertisements tailored to each customer's unique needs.

8. Intelligent Government Administration and Services

Cities worldwide are deploying various technologies to create "smart cities," where buildings, other infrastructure, and people are all connected to ensure that everyone and everything moves as safely and smoothly as possible. A city, for example, may collect and evaluate endpoint data on traffic flow and then utilize the results to direct drivers away from crowded regions. A system with many moving parts necessitates 5G, driving significant investment in this area.

9. Utilities That Are More Efficient and Effective

Smart grids rely on a large IoT ecosystem, which includes many endpoint sensors, edge devices, and analytics capabilities scattered across large geographic areas, which necessitates the kind of connectivity that 5G provides.

According to Jefferson Wang, global 5G head at Accenture, "a genuinely smart grid unlocks several benefits, including enabling utilities to cut operational costs and increase the robustness of energy services." A utility, for example, can utilize sensors to monitor the risk of wildfires in real-time, allowing for better prevention and faster response times in the event of an actual catastrophe. This can then assist utilities in avoiding service interruptions and preventing serious infrastructure damage.

10. More Robust Workforce Assistance

5G is going to have a significant impact on workers in a variety of industries. It can assist remote work by providing a significantly quicker and more stable connection with adequate robustness to handle even mission-critical distant work. It also allows for distant cooperation through the use of AR and other modern technologies. It also encourages the use of AR and related tools in training.

5G also helps advanced workplace safety programs. For example, a manufacturer can utilize a video analytics system with 5G connectivity to assess and respond to significant safety issues, such as preventing equipment from starting if the system identifies a worker who is not wearing safety equipment.

CHAPTER 3

Security in the 5G Era

Several standard organizations and working groups are working on the 5G specifications (Figure 3.1). Here, we are trying to overview the most important ones.

5G SECURITY

FIGURE 3.1   Standardization organizations of relevance for 5G security.

THE 3ᴿᴰ GENERATION PARTNERSHIP PROJECT (3GPP)

3GPP brings together numerous regional standard-setting organizations in telecommunications to produce and maintain worldwide technical specifications. Cellular telecommunications technologies such as radio access, core network and services, non-radio access to the core network, and interworking with non-3GPP networks are included in the scope.

3GPP SA312 is the most active working group tackling 5G security and privacy issues. The group is in charge of determining the security and privacy requirements and establishing the security architectures and

DOI: 10.1201/9781003264408-3

protocols that will be used to meet them. Moreover, 3GPP SA3 assures that the cryptographic method required for 5G security specifications is available.

THE EUROPEAN TELECOMMUNICATIONS STANDARDS INSTITUTE (ETSI)

ETSI has several subgroups dealing with different elements related to 5G system security (Figure 3.2).

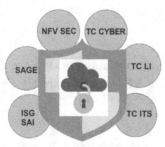

ETSI 5G SECURITY

FIGURE 3.2   ETSI working groups.

Network Function Virtualization Security Working Group (NFV SEC)

On NFV security challenges, the NFV SEC working group has advised the ETSI Industry Specification Group NFV (ETSI ISG NFV)18. Network, computer, and cloud security specialists from operators, suppliers, and law enforcement organizations are among the attendees.

Technical Committee for Cybersecurity (TC CYBER)

ETSI TC CYBER has developed specifications for protecting the Internet.

Technical Committee for Lawful Interception (TC LI)

ETSI TC LI has developed standards to support the technical requirements of law enforcement.

Technical Committee for Intelligent Transport Systems (TC ITS)

The ETSI TC ITS Working Group 5 (WG5) is concentrating on the security concerns of Cooperative ITS (C-ITS), which involves vehicles communicating with one another and/or with the transportation infrastructure. The security framework's significant aspect has been designing and developing a security management infrastructure for C-ITS.

Industry Specification Group on Securing Artificial Intelligence (ISG SAI)

The ETSI ISG SAI is a new organization (it held its first meeting in October 2019) that intends to develop technical standards to improve Artificial Intelligence (AI) security. Securing AI from attacks, mitigating hostile AI, and using AI to improve security measures are key research areas.

Security Algorithms Group of Experts (SAGE)

ETSI SAGE has been working on cryptographic algorithms and protocols to prevent fraud and unwanted access to public and private telecommunications networks and to protect user data privacy. The group has developed encryption, authentication, and key generation methods for numerous mobile technologies, including 2G/3G/4G32. More recently, the group has worked with 3GPP to develop cryptographic algorithms for 5G, such as the support offered for 3GPP and Study on the Support of 256-bit Algorithms for 5G.

ITU (INTERNATIONAL MOBILE TELECOMMUNICATIONS) TELECOMMUNICATION STANDARDIZATION SECTOR (ITU-T)

Various technical Study Groups (SGs) of the ITU-T work on standardization, where ITU-T member members generate consensus-based Recommendations (standards). The ITU-T X-series Recommendations address OSI, information and network security, cyberspace, cloud computing, and data security.

The ITU-T SG 17 is in charge of addressing the security of 5G systems. The group's 5G security initiatives include creating X-series Recommendations on the following topics.

- Software-Defined Network (SDN);

- NFV;

- Internet of Things (IoT)

- Big data analytics in mobile Internet services;

- Cloud computing; and

- Cryptographic profiles

THE INTERNET ENGINEERING TASK FORCE (IETF)

The IETF is an Internet standards body that deals with protocols and architectures for delay-sensitive and delay-tolerant applications, as well as IP layer, network management, routing, end-to-end data transmission on the Internet, and security.

Other Stakeholders

The GSM Association, or GSMA, is one of the most critical stakeholders in the field of mobile communications.

The GSMA maintains a security division that helps its members defend their mobile systems. The team collaborates with the GSMA Fraud and Security Group (FASG), the GSMA's leading group dealing with 5G-related fraud and security issues.

GUIDING PRINCIPLES TO UNDERSTAND SECURITY STANDARDIZATION FOR 5G AND BEYOND

The following are the guiding principles.

- The backbone of mobile network security is standardization—standards are open and universally agreed upon, ensuring interoperability and transparency.

- There is no single security standard covering everything—5G contains various standards from various standardization groups.

- Security isn't only about standardization; it's also about the network's implementation, deployment, configuration, and operation.

- Security assurance satisfies higher security requirements through established auditing methods in which items are compared to standardized specifications.

To give direction to cyber security organizations, the 3GPP has created 5G standards, including methods for encryption, mutual authentication, integrity protection, privacy, and network availability. The standards, according to 5G Americas, a trade organization for mobile operators, include:

- A single authentication system that supports concurrent connections and allows for smooth mobility across multiple access technologies.

- Protection of user privacy for sensitive information that is frequently used to identify and track subscribers

- Slice isolation and Secure Service Based Architecture (SBA) optimize security by preventing vulnerabilities from spreading to other network slices.

- SS7 and diameter protocols for roaming are being improved.

- Adding native support for secure steering of roaming (SoR) allows operators to direct users to preferred partner networks, enhancing customer experience, lowering roaming rates, and preventing fraud.

- Improved identification and mitigation mechanisms for rogue base stations

- In addition, proprietary operator and vendor analytics systems provide additional security levels.

IMPORTANCE OF 5G SECURITY

Organizations with a digital presence and a telecommunications investment have traditionally prioritized cybersecurity in the cyber world. When advances in technology such as virtualization, IoT, SDN, NFV, edge computing, and Industry 4.0 are combined with equally broad but deteriorating cybersecurity, the security and functionality of these environments will be significantly impacted. Between the radio access network (RAN) and the core network, several 5G protocols are flexible enough to overlap various types of physical and virtual elements. Separating the RAN and core network functions in the telecommunications environment will undoubtedly affect performance, but it will also have security implications, including SDN, NFV, and edge computing difficulties.

In the information technology landscape, 5G security concerns can lead to lower traffic visibility, whereas a lack of WAN (Wide Area Network) solutions such as Secure Access Service Edge (SASE) can lead to some business traffic visibility. Because the rise of 5G and its ability to connect to a large number of devices is linked to rising IoT usage, the security issues of the latter also influence the former. Because IoT devices are notoriously insecure, the vulnerability is expected to spread to the organization's security framework as IoT and related 5G technology become more prevalent. As a result, business organizations must implement IoT security solutions to safeguard the safety of their devices.

Similarly, the limited 5G supply chains will cause security difficulties because new mobile technologies are more software dependent than old mobile networking, increasing the attack surface. The telecom network is equally important when conceptualizing security, and 5G security involves understanding the aspects such as:

- Increased stake value
- Risk tolerance
- Physical & virtual dependencies
- Security standards, protocols, deployments, and operations
- Proactive cybersecurity measures
- Vulnerability management
- Supply chain security

5G-ENSURE ARCHITECTURE

5G-ENSURE develops and delivers a 5G reference security architecture that is shared and agreed upon by diverse 5G stakeholders and is supported with usable security enablers that solve essential concerns.

The 5G-ENSURE Security Design focuses on logical and functional architecture while ignoring (most) aspects of physical and deployment architecture. General developments such as network deperimeterization and 5G systems' strong reliance on software-defined networking and virtualization, in general, have prompted this attention. The security architecture is expanding based on the current 3GPP security architecture.

The 5G-ENSURE architecture expands and revises the 3GPP security architecture from TS 33.401 to combine essential features and the domain idea from 3GPP TS 23.101 with supporting trust models for a 5G vision that goes beyond "telecom" and "mobile broadband".

- Infrastructure domains and tenant domains to capture the physical and logical aspects
- Management domains to capture orchestration and security management
- Identity Management (IM) domains to re-use existing industrial AAA for device authentication

- Internet Protocol (IP) domains to model external IP networks

- Slice domains to capture network slicing, application domains transversal to others

The 5G-ENSURE architecture's logical "dimension" captures, first and foremost, the security aspects related to the many domains involved in delivering services over 5G networks. As a result, this section is closely linked to the project's trust model. Security aspects related to network tiers and/or unique types of network traffic are also captured in the logical section.

The architecture includes security capabilities required to safeguard and maintain the security of the various domains and strata. A stratum is a collection of protocols, data, and functions connected to a certain component of one or more domains' services. The following strata are depicted in Figure 3.3.

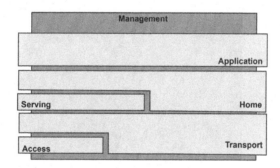

FIGURE 3.3   5G ENSURE stratum.

- The **Application Stratum** represents the application process itself, provided to the end-user.

- The protocols and functions relevant to managing and storing subscription data and home network-specific services are found in the **Home Stratum**.

- The Serving Stratum is a sub-stratum of the Home Stratum that includes protocols and functionalities for routing and forwarding data/information created by users or networks from source to destination.

- The **Transport Stratum** supports user data transport and network control signaling from other strata through the network.

- The **Access Stratum** is a sub-stratum of the transport stratum.

- Aspects of traditional network management (configuration, software upgrades, user account management, log collection/analysis) and, in particular, security management are included in the **Management Stratum** (security monitoring audit, key and certificate management, etc.). This stratum contains aspects of virtualization administration and service creation/composition (orchestration, network slice management, isolation and VM (Virtual Machine) management, and so on).

CHAPTER 4

Security Standards and Their Role in 5G

There are various security standards available for 5G. In detail, we are discussing security standards like 3ʳᵈ Generation Partnership Project (3GPP), a Service-based concept, and Network slicing security and protection at the Network or transport layer.

3GPP 5G SECURITY STANDARDS

3GPP is a consortium of seven telecommunications standard development organizations collaborating to produce the reports and specifications defining 3GPP technologies. The project includes work on codecs, security, and quality of service, as well as cellular telecommunications network technologies such as radio access, the core transport network, and service capabilities. As a result, 3GPP provides entire system specifications, including hooks for non-radio core network access and Wi-Fi network interworking.

It specifies the 5G system's and 5G core's security architecture, features, and methods. It also covers the security processes used across the 5G system, including the 5G core and 5G New Radio.

The following security domains are significant in 5G systems:

- **Network access security (I):** It is a set of security characteristics that enable a user equipment (UE) to securely authenticate and access network (AN) services, including 3GPP and non-3GPP access, as well as to guard against attacks on (radio) interfaces. It also comprises the transfer of security context from SN (secondary node) to AN for access security.

DOI: 10.1201/9781003264408-4 **41**

- **Network domain security (II):** A set of security features that allows network nodes to securely communicate signaling data and user plane (UP) data.

- **User domain security (III):** It is a collection of security elements that protect a user's access to mobile devices.

- **Application domain security (IV):** A set of security features allows user and provider domain apps to exchange messages securely.

- **Service-based architecture (SBA) domain security (V):** A set of security characteristics that allow the SBA architecture's Network functions to securely connect within the serving network domain and other network domains. Network function registration, discovery, authorization security aspects, and protection for service-based interfaces (SBI) are examples of such characteristics.

- **Security visibility and configurability (VI):** A set of features that allow the user to know whether a security feature is active or not.

SOME PROMINENT FEATURES DEFINED FOR 5G SECURITY BY 3GPP

Following are some of the features described for 5G Security:

Increased Home Control

Home control is utilized to authenticate the device's position when the device is roaming. When the home network receives a request from a visiting network, it can check to see if the device is indeed in the Network.

Home control was added to address vulnerabilities discovered in 3G and 4G networks, where networks might be faked by sending bogus signaling messages to the home network requesting the device's International Mobile Subscriber Identity (IMSI) and location. As a result, this data might be utilized to listen in on phone calls and read text messages.

Unified Authentication Framework

Both 3GPP and non-3GPP ANs employ the same authentication techniques (for example, 5G radio and Wi-Fi access).

Extensible Authentication Protocol (EAP) native compatibility allows for future plug-in authentication methods to be introduced without affecting the serving networks.

Security Anchor Function (SEAF)

With the new SEAF in 5G, the concept of an anchor key is introduced). When a device moves between multiple ANs or serving networks, the SEAF allows it to re-authenticate without running the complete authentication process (for example, Authentication and Key Agreement [AKA]). This decreases the signaling strain on the home network's Home Subscriber Server (HSS) during various mobility services. Separation or co-location of the SEAF with the Access and Mobility Management Function (AMF) is possible. The SEAF capability is co-located with the AMF in 3GPP Release 15.

Subscriber Identifier Privacy

Each subscriber in 5G is assigned a globally unique Subscriber Permanent Identifier (SUPI). IMSI and Network Access Identifier (NAI) are two examples of SUPI forms. When a mobile device establishes a connection, the SUPI is never revealed over the air in the clear. This is in contrast to 3G and 4G networks, where the IMSI is revealed during the attach operation (the third vulnerability in 3G and 4G networks) before the device can even authenticate with the new Network.

Until the device and Network are authenticated, a Subscription Concealed Identifier (SUCI) is utilized instead of disclosing the SUPI. The SUPI is only then disclosed to the serving network by the home network. This process was created to prevent IMSI catchers (also known as false base stations or Stingrays) from obtaining the subscriber's identity. This is performed by compelling a device to connect to either the Rogue Base Station (RBS) or the operator's Base Station while sniffing unencrypted traffic over the air.

Security Edge Protection Proxy (SEPP)

The 5G system design implements a SEPP at the Public Land Mobile Network (PLMN) perimeter to protect messages sent through the N32 interface. SEPP receives and protects all service layer messages from the Network Function (NF) before sending them out over the N32 interface. Additionally, after checking security, it gets all messages on the N32 interface and transmits them to the relevant network function.

For all layer information sent between two NFs via two distinct PLMNs, the SEPP implements application-layer security (Figure 4.1).

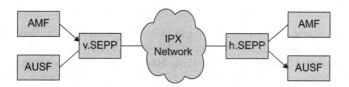

FIGURE 4.1   The role of the SEPP in the security architecture.

However, new/improved techniques to tackle threats within a service (vertical) or across a set of services (horizontal) are accessible in terms of secure 5G deployment considerations. This would include the following for horizontal, system-wide security (Table 4.1):

TABLE 4.1 Security Requirements for 5G Network Elements

5G Network Elements	Security Requirements
UE	– Encryption of signaling and user data between UE and gNB for reasons of confidentiality. – Ensuring data integrity for signaling and user data between UE and gNB – Secure storage and processing of the login information from the user Profile – Protection of privacy through encryption and secure storage of keys in the USIM – Calculation of the SUCI (Suscription Concealed Identifier)
gNB	– Encryption of signalling and user data between UE and gNB for reasons of confidentiality – Ensuring data integrity for signaling and user data between UE and gNB – Authenticating and authorizing a gNB during setup and configuration – Protection of the gNB software – Protection of the keys used and stored in the gNB – Secure processing and storage of user and signaling data – Providing a secure environment for all sensitive data – Secured transmission on F1 interface when splitting a gNB into CU and DU – Secure transmission on E1 interface when dividing a CU into CU-CP and CU-UP
AMF (Access and Mobility Management Function)	– Because of the confidentiality encryption of NAS signaling – Ensuring data integrity for NAS signaling – Triggers the primary authentication of the UE via SUCI (Subscription Concealed Identifier)
SEAF (Security Anchor Function)	– Provides the authentication function via the AMF in the Serving Network – Supports the primary authentication of the UE
UDM (Unified Data Management)	– Long-term keys for authentication and security association must be protected and must not leave the UDM/ARPF environment (Authentication credential Repository and Processing Function) – Provides SIDF service

UE—User equipment, AMF—Access and mobility management function, USIM—Universal Subscriber Identity Module, SUCI —Subscription Concealed Indentifier, DU—Distributed Unit, UDM—Unified Data Management, ARPF—Authentication credential Repository and Processing Function, SIDF—Subscription identifier de-concealing function

- Increasing the Network's resiliency

- Network slicing and security features based on needs

- Application-level security that makes use of other domains' trust stack

- Protection of confidentiality and integrity across the radio network

- Transport Layer Security (TLS) between 5G Core operations, independent of architecture

- (SBA) permits functional-level components to be divided even at the radio unit level. Hardening the virtualized stack and using trusted layers within embedded systems are vital for increased vertical security across all functional parts. This may require the virtualized layers to utilize trusted hardware components (through trusted platform modules [TPM], hardware security modules [HSM], or secure enclaves) and expose them vertically to apps.

SYSTEM-WIDE SECURITY (HORIZONTAL SECURITY)

Security across the entire system Consumers and businesses use existing (4/3/2G) cellular networks for mobile broadband (connectivity services), messaging (e.g., SMS), and telephony services, as previously stated. The reliance on cellular networks to provide reliable and secure communication is growing as societal behavior and corporate services evolve.

5G's goal is to become a dependable and trusted innovation platform for businesses and organizations to design and deliver new added-value services, but it's also seen as a key facilitator for digitizing and updating important national infrastructures like electricity and transportation. The latter improves the threshold for 5G systems, ensuring higher availability and improved security assurances. The horizontal, system-wide security strategy stretches across the Network from the user device to the reference point where the operator terminates their services.

Horizontal security is achieved by combining and synchronizing a variety of security controls across various domains in telecommunication networks, such as radio access (e.g., radio units, baseband units, antennas), transport networks (e.g., optical equipment, Ethernet bridges, Internet Protocol (IP)/ MPLS routers, SDN controller), packet core (e.g., MME, S-GW, PGW, HSS), network support services (e.g., DNS, DHCP [e.g., network management, customer experience management, security management]). To enable the targeted availability of services as well as the confidentiality and integrity of data sent, saved, and processed within the 5G system, security across all of these domains must be coordinated. Horizontal security will also protect 5G users' privacy by ensuring that data transported through the system is always secure in terms of confidentiality and integrity.

Network-Level

Transport networks are critical components of the 5G system because they allow high-speed, low-latency communication between all 5G network activities. As a result, the availability of transportation networks is inversely proportional to the availability of the 5G system and the services it offers.

Transport networks can use a variety of technical solutions as well as considerations during network design to ensure the availability of transport services during node failure, cable or fiber breaks, or overload events, such as:

- Geo-redundant paths allow traffic to be rerouted in the event of a path failure

- Link redundancy solutions that allow for quick failover in the event of a port or link failure

- Path redundancy mechanisms, which reroute traffic flows in the event of a failure or overload

- Configure important network nodes to be high-availability to manage node failure

- Traffic segmentation mechanisms (e.g., VLAN and MPLS) are used to divide the traffic between domains logically

- Quality of Service (QoS) enforcement for resource and congestion management, including traffic queuing methods, rate restriction, and traffic policing.

- Detection and mitigation of distributed denial-of-service attacks (DDoS).

- Authentication based on ports to ensure that only authorized network devices are connected to the Network

- IPsec or MACsec to construct encrypted and authenticated tunnels for data transfer between sites and network elements

Protection at the Network or Transport Layer

Client and server certificates, as well as TLS, which is intended to be used for transport protection, will be supported by NFs, while NDS/IP will continue to be used for network layer protection.

When considering the request for employing TLS, it's worth noting that while "all NFs shall support TLS," the actual use of TLS for transport security does not appear to be as clearly specified. Instead, "TLS shall be utilized for transport protection within a PLMN unless network security is supplied by other means." If "by other means" has lower protection than TLS, this could open the door for flaws to be introduced.

Additionally, in the case of trustworthy (e.g., physically protected) interfaces, it is up to the operator to decide whether or not to utilize cryptographic protection. Depending on the extent of physical protection provided, this may increase the danger to data confidentiality.

If SBIs are not adequately protected at the transport layer, sensitive information/data may be exposed and later interfered with.

Network Slicing Security

Network slicing is gradually becoming one of the most distinguishing characteristics of today's 5G networks. In a nutshell, network slicing refers to a network's capacity to autonomously build and execute numerous logical networks as practically independent business activities on shared physical infrastructure. Slicing is projected to be a key component of the 5G network's design, as it will support the bulk of 5G use cases.

A network slice, according to 3GPP specifications, is a complete logical network (offering Telecommunication services and network capabilities) that includes the Access Network (AN) and the Core Network (CN). Multiple logical networks can be managed as essentially independent business processes on a single physical infrastructure, thanks to network slicing. In practice, this translates to the idea that the mobile Network could be divided into a series of virtual resources. Each one of them is referred to as a "slice," and it can be used for many purposes. A slice could be assigned to a mobile virtual network operator (MVNO), an enterprise customer, an Internet of Things (IoT) domain, or any other group of services that makes sense (for example, mobility as a service). The access point name (APN) idea utilized in today's mobile Network is extended by a network slice.

Network slicing is a technique for comparing the end-to-end performance of one network segment to another. Slicing makes it possible to scale a large number of solutions.

Security isolation is done by isolating both physical and logical networks properly. Operators can use numerous logical networks to configure the physical Network. Multiple logical networks with varied network capabilities can also be configured by operators (e.g., namespaces, VLANs). Each network slice in the 5G environment can support one application or a group of applications, each of which can deliver one or a mix of services (e.g., voice communication, video streaming, or IoT). Traffic and resource isolation, in addition to physical and logical segregation, is crucial. The creation of specific virtual lanes to service various traffic types that can be routed through a specific logical/physical lane pair to offer traffic isolation. The underlying architecture should support resource isolation. Isolating traffic and resources is essential for protecting vital information systems from security flaws and attacks. For example, if a particular lane is penetrated, the scope of a prospective attack can be limited to that slice.

Note: Although virtual slices can be an excellent example of isolation, it's important to remember that the slice numbers, the amount of traffic and resources that go into each slice, and the slice security setup must all be suited to the individual use case(s) being served.

When slicing is used, it supports the following aiding functions:

- Physical Resource Separation

  - Over the air via 5GqoS—Different channels or at resource block level

  - Over the air—with a dedicated spectrum (for example, dedicated safety of public)

  - Core Network Element functions—dedicated elements of a network are reserved for particular slices

- **Traffic separation:** Both in the RAN and of mappings in transit networks (e.g., VLANs, MPLS, tunnels).

- **Custom traffic routing/steering:** to meet the needs of end users. Virtualization of the 5G core is accompanied by virtualized networks, allowing for easy connection with other cloud environments and steering and VPN features for enterprise customers. Service chaining can also be done with the steering.

- **Scalability:** Many of these functions were covered in the past based on APNs. It is now termed DNNs in 5G. Adding of slices greatly enhances the scalability. In addition, private networks that work in 3GPP have created the Network Identifier (NID), similar to an APN/DNN but based on the NIA type, which has unlimited scale. Slice also inherit this scale as they can be defined/managed per NID and per DNN.

- **Optimization:** The 5G standards also suggest more formality and standardization of optimization in a Network (For example, caching of per UE per NID/DNN slices in AMF)

Application-Level Security

For all service layer information sent between two NFs via two distinct PLMNs, the SEPP implements application-layer security.

All the IEs in the Hyper Text Transfer Protocol (HTTP) message payload, crucial information in the HTTP message header, and the request URI make up the application layer traffic. In SEPP, not all IEs are treated equally in terms of security. End-to-end (E2E) encryption is required by some IEs, whereas others require E2E integrity protection. Others may want E2E integrity protection that is changeable while in transit by an intermediate Internetwork Packet Exchange (IPX) provider.

To allow trustworthy intermediary IPX nodes to see and alter certain IEs in HTTP messages while still securing all sensitive information between SEPPs end-to-end, the SEPP implements application layer security in such a way that:

- Sensitive data such as authentication vectors are fully E2E and safeguarded by two SEPPs. This assures that no IPX network node can see such data while it is enroute.

- IEs that are subject to modification by intermediary IPX nodes are integrity protected and can only be edited by authorized IPX nodes in a verifiable way.

- SEPP receivers can identify unlawful IPX node modifications.

The SEPP must comply with the following criteria:

- Serve as an opaque proxy node

- Encrypt application-layer control plane messages sent between two NFs from separate PLMNs that interact using the N32 interface

- Use the roaming network's SEPP to perform mutual authentication and cipher suite negotiation

- Manage the key management portions of the N32 interface between two SEPPs, including setting up the requisite cryptographic keys for safeguarding messages

- Hide internal topological information from external parties by performing topology hiding

- As a reverse proxy, provide internal NFs with a single point of access and control

- Confirm that the sending SEPP has permission to use the PLMN ID in the received N32 message as the receiving SEPP

- Make a clear distinction between certificates used for peer SEPP authentication and certificates used for message modification intermediates.

- N32 signaling messages that are faulty should be discarded

- Enable rate-limiting functionality to protect itself and subsequent NFs from excessive CP signaling, which includes SEPP-to-SEPP signaling messages.

- Enable cross-layer validation of source and destination addresses and identities by implementing anti-spoofing methods (for example, FQDNs or PLMN IDs)

- TLS shall be used between the SEPPs if there are no IPX entities between them. If there are IPX entities between SEPPs, application-layer security on the N32 interface between SEPPs is required to ensure that the SEPPs are protected.

CONFIDENTIALITY AND INTEGRITY PROTECTION

Data confidentiality protects data against unintentional, unlawful, or unauthorized access, disclosure, or theft. Generally, integrity protection refers to mechanisms that protect particular software's logic and/or data. Some mechanism is included in the 5G architecture for confidentiality and integrity protection.

Protection of User and Signaling Data

Using strong encryption to safeguard the radio path between the UE and the base station is not new. Encryption of traffic between mobile stations and eNodeB can be implemented in contemporary 4G networks to safeguard the confidentiality and integrity of data transferred between the

UE and the Mobility Management Entity (MME). Similar considerations apply in 5G networks. Some of these criteria are optional, as they were in 4G. The preservation of UP data integrity is a new feature of 5G, but it is also optional.

Confidentiality Protection

User and signaling data confidentiality requirements to be implemented on the UE. In terms of user data and RRC signaling, these requirements are effectively matched for the gNB and AMF in terms of NAS signaling. User and signaling data confidentiality protection is an option in all circumstances.

Four ciphering algorithms can be used to protect confidentiality in this way. Following are the algorithms:

1. NEA0—no encryption, plaintext (therefore offering no protection)

2. SNOW 3G cypher (128-NEA1) allows for backward compatibility with 3G networks

3. 128-NEA2 is an AES-128 CTR cipher that allows for backward compatibility with 4G LTE networks

4. 128-NEA3 is a 5G-specific stream cipher based on the ZUC stream cypher

The UE must support the first three algorithms (NEA0, 128-NEA1, and 128-NEA2). In contrast, support for the fourth algorithm (128-NEA3) is optional.

According to the specification, secrecy protection should be used where the operating territory's regulations allow. In other words, unless specified local conditions forbid it, the default security setting is always enabled.

In terms of security issues, user data is exposed to interception due to a lack of confidentiality protection. Similarly, a lack of confidentiality protection for signaling data may allow attackers to intercept status and authorization data between the UE and the gNB/AMF, tracking user position and executing other passive or active attacks.

Integrity Protection

Control plane communication is not only protected for privacy by applying encryption but also protected against malicious modifications possibly applied by an attacker. This type of protection is termed "integrity protection"—assuring that a legitimate source indeed generated the data reaching the destination and that no modifications were applied enroute.

Integrity Protection of the User Plane

The UP integrity protection between the device and the gNB was included as a new feature in 5G. The integrity protection feature, like the encryption feature, is required on both devices and the gNB, although its use is optional and under the operator's authority.

Integrity protection is well acknowledged to be resource-intensive, and not all devices will be able to handle it at the maximum data rate. As a result, the 5G System enables negotiation of which rates are appropriate for the feature. For example, suppose a device's maximum data rate for integrity-protected traffic is 64 kbps. In that case, the Network only enables integrity protection for UP connections when data rates are not expected to exceed 64 kbps.

User and signaling data integrity, as well as replay protection, are required on the UE. In terms of user data and RRC signaling, these requirements are effectively matched for the gNB and AMF in terms of NAS signaling.

Integrity protection is obligatory for the signaling plane (RRC-signaling and NAS-signaling) but optional for the UP based on these requirements.

Four integrity protection algorithms can be employed for this purpose. These are the algorithms:

1. NIA0—no encryption, plaintext (therefore offering no protection)

2. Based on SNOW 3G, 128-NIA1

3. 128-NIA2—in CMAC mode, based on AES-128

4. Based on 128-bit ZUC, 128-NIA3

The UE must support the first three algorithms (NIA0, 128-NIA1, and 128-NIA2), whereas support for the fourth algorithm (128-NIA3) is optional, according to the standard.

The lack of integrity and reply protection in signaling data traffic between the UE and the gNB or AMF can lead to data compromise and tampering and assist different man-in-the-middle attacks. Furthermore, as a new feature is added to the 5G specification, the ability to have UP integrity protection is critical to avoid a malicious change of user data. As studies have shown, such alterations can have a significant impact, such as routing DNS requests from the UE to a malicious server.

Note: Because integrity protection is a resource-intensive function, not all devices may be able to support it all of the time or only at specific rates. When older equipment lacking support for UP integrity protection is employed in 5G NSA situations, further concerns may develop. The 3GPP technical paper TR 33.85385 delves more into some of these challenges and provides remedies.

There is also an optional PDCP COUNT process that the gNB can utilize to do a local authentication and check the quantity of data exchanged between the gNB and the UE both up-and down-stream regularly. Any disparity in the count tally will indicate maliciously inserted packets; hence the check can be used to detect any tampering or intervention between the two devices. Such a count check is redundant when integrity protection is enabled. When no integrity protection is engaged, utilizing such a counter may provide some amount of integrity protection with a low overhead to the operator and user. Still, such control is not deemed to be a sufficiently safe replacement for the cryptographic integrity protection mentioned above.

User Plane Integrity Protection—Man-in-the-Middle
The Access Stratum over-the-air flows in 5G, like 4G, are afflicted by a lack of UP Integrity Protection (IP). Although the 3GPP specifies UP IP integrity techniques, many OEMs (Original Equipment Manufacturers) have chosen not to implement them due to the negative impact on the user experience (e.g. download and upload data throughputs). Enabling UP IP necessitates a significant amount of computational resources and adds overhead, both of which have a direct impact on the maximum throughputs that can be measured on the user device. Although IP is allowed on Control Plane communications, the user's data traffic is still susceptible since the Control Plane and UP are separated. A rogue base station, for example, could leverage the lack of UP IP to modify user data messages (i.e. DNS) and send a user to a malicious website. In 5G, the 3GPP SA3 Security Working Group is considering UP IP requirements that enable OEMs to implement capabilities in gNB radios and 5G baseband modems for the user equipment. To reduce the negative effects on the user's download and upload data experience, several 5G radio OEMs are creating hardware-based encryption acceleration.

SERVICE-BASED ARCHITECTURE (SBA)

5G has ushered in a paradigm shift in mobile network architecture, moving away from the traditional concept of point-to-point network function interfaces and toward SBIs.

The 3GPP-defined 5G core network is a SBA in which a 5G network's control plane functions and common data repositories are delivered by a group of interconnected NFs, each having authority to access the services of others.

NFs are self-contained, self-contained, and reusable components. Each NF service exposes its capabilities using a SBI based on a well-defined RESTful HTTP/2 interface. The Quick UDP Internet Connections (QUIC) protocol may be used in the future to address TCP head-of-line (HOL) blocking difficulties.

SBA provides for separating functional-level components even at the radio unit. Hardening the virtualized stack and using trusted layers within embedded systems are vital for increased vertical security across all functional parts. This may require the virtualized layers to utilize trusted hardware components (through TPM, HSM, or secure enclaves) and expose them vertically to apps.

Service-Based Architecture (SBA) and Interconnect Security

5G has ushered in a paradigm shift in mobile network architecture, moving away from the traditional concept of point-to-point network function interfaces and toward SBI. The many capabilities of a network entity are refactored into services exposed and offered on-demand to other network entities in a SBA.

In addition to protecting communication between core network entities at the internet protocol (IP) layer, the adoption of SBA has pushed for protection at higher protocol layers (i.e., transport and application) (typically by IPsec). As a result, the 5G core network functions support cutting-edge security protocols like TLS 1.2 and 1.3 at the transport layer to protect communication and the OAuth 2.0 framework at the application layer to ensure that only authorized network functions have access to a service provided by another function.

Three building pieces make up 3GPP SA3's improvement to interconnect security (i.e., security between various operator networks):

- In the 5G design, a new network function called SEPP was implemented. These security proxies are expected to carry all signaling communications across operator networks.

- The authentication of SEPPs is also necessary. This allows for effective traffic filtering from the connection.

- Finally, on the N32 interface between the SEPPs, a new application layer security solution was built to safeguard sensitive data properties while still allowing mediation services throughout the interconnect.

The primary components of SBA security are authentication and transport protection across network services using TLS, authorization framework using OAuth2, and improved interconnect security utilizing a new security protocol designed by 3GPP.

TABLE 4.2   Layered 5G security controls

Layered 5G Security Controls	
Orchestration	Securing Orchestration management & interfaces, Securing Policy Enforcement, and enhancing visibility within Orchestration and between Orchestration and network components
User	Segmentation, User Access based on Zero Trust principles, DNS protection
Network	Segmentation, Policy enforcement, Securing Network interfaces, Securing Cloud integration, and workloads, Securing peering & Roaming interface
Application	Securing 3rd Party application interfaces, DDoS protection, Application security, DevSecOps practices, Segmentation, Cloud application policy sync, and enforcement, securing API
VNFs and CNFs	Securing VNF/CNF, securing Software Lifecycle, Isolation between VNFs/CNFs, detecting malicious virtual functions and vulnerabilities
Infrastructure	Hardening of NFVI, perimeter security, DDoS protection, securing-E-W traffic

(CNF)—Cloud-Native Network Functions, VNFs—Virtual network functions, NFVi (network functions virtualization infrastructure), (DDoS)-Distributed denial of service, DNS(Domain Name System)

5G FUNCTION ELEMENT DEPLOYMENTS (VERTICAL SECURITY)

One of the most critical issues is how the new 5G core's network operations will be delivered, and virtualization is vital for enabling flexible deployments.

The 3GPP standard outlines network functions and how they interact, but not how they should be implemented in embedded systems or virtual environments. Traditionally, vendor-designed hardware platforms have been used to construct radio base station equipment and radio core nodes. These platforms were meticulously engineered to meet stringent requirements in terms of availability, mean time between failures (MTBF), performance, scalability, power consumption, and physical security.

A radio baseband unit, for example, may feature tamper-resistant hardware for securely storing sensitive secrets, secure boot methods for verifying the integrity and origin of software put into the hardware, and hardware accelerators to improve cryptographic performance. The hardware is furnished with unique vendor credentials, known as Vendor Credentials, that are used to cryptographically identify the device vendor of origin during the production of baseband units. This credential is used to secure baseband unit deployment and integration into operator networks. These credentials are securely saved on hardware devices with a Trusted Execution Environment (TEE) as defined by the Trusted Computing Group, resulting in hardware-rooted trust that may be carried into deployment.

Multiple manufacturers may deliver different aspects of the solution, including the hardware infrastructure, the virtualization platform, and the applications that perform the 3GPP network functions in virtualized deployments. To enable a secure deployment in virtualized deployments, secure provisioning and storing of IDs and credentials is critical. Currently, the industry is attempting to develop techniques for achieving similar levels of trust and security to those found in embedded systems. For example, TPM, HSM, and secure enclaves in CPUs must be included in hardware platforms (data center servers). Moreover, these capabilities must be used by the virtualization platform and exposed to and attested by applications running on those platforms. Virtualization of 3GPP functions enables for more flexible deployment of services across network infrastructure than is achievable with hardware-based solutions.

It is feasible, for example, to construct a network slice on a distributed cloud platform that serves local enterprise services or regional IoT applications by deploying both RAN and core network operations deeper in the Network near the edge. This necessitates the hardening and enforcement of required security controls by network orchestrators that deploy the apps, the distributed cloud platform on which the applications run, and the applications themselves. This is necessary to satisfy both the operator's desired security posture and the security need for the network slice use case. This is accomplished with the help of systems that coordinate service deployment and security setup across all domains concerned. Security monitoring is required after deployment to ensure that the desired security state is maintained throughout the lifecycle of deployed services.

THREATS, VULNERABILITIES, AND ATTACKS—5G STANDALONE

Furthermore, the commercialization of Massive Machine-Type Communications (mMTC) and Ultra-Reliable Low Latency Communications (URLLC) use cases will be enabled by the 5G Core Network and the implementation of 3GPP Release 16 requirements.

Areas of improved wireless privacy that mask the IMSI, an SBA (leveraging Cloud-Native methods,) SDN, and Network function virtualization (NFV) are major features that distinguish 5G SA from 5G NSA and older generations of wireless networks.

Service Based Architecture (SBA)

The 5G ecosystem is mostly made up of software that can run on general-purpose hardware and connects with APIs. The program's integrity, particularly from open-source locations, as well as the broader software supply chain, is a subject of concern. 5G is built on Cloud-Native principles, which allow services to be created, deleted, and communicated with one another in real-time. To avoid illegal instructions or access to resources, all systems must be adequately authenticated using protected communication (having the ability to instruct the system to exhaust its resources is one form of DoS attack). Users will have access to network-specific services under the 5G architecture.

Any existing hardware or software flaws (including operating systems) will be present in the 5G architecture. SDN and NFV are two concepts that will be used in 5G, and each has its own set of risks and weaknesses.

SDN (Software Defined Networking)

5G systems and functions are programmable software modules at their core. Understanding the security risks of programmable modules in large systems has taken a lot of time and effort. The capacity to program functionality means that the operator, or even the user, can alter the system's overall behavior or software. As a result, only authorized entities should be able to update or program the Network, and the provider should vet and regulate the end capabilities. Both human-selected and automated service chains are affected. If programmability is available, there must be a mechanism to test whether changes in behavior create the desired result rather than unexpected consequences prior to deployment. Currently, network verification techniques are employed to ensure that a network follows its intended policies. It's unclear whether the network verification mechanism itself may be attacked, leading the operator to believe guidelines are being followed when they aren't.

Network Function virtualization (NFV)

The integrity of the code that makes up a virtualized function and the interaction between virtualized functions are both critical. In any setting where open-source software can be employed, it is a source of risk. The 5G system's operations can be made up of open-source software components, but their security and integrity aren't always guaranteed. In a standardized, API-style environment, the virtualized pieces must communicate with one another. The APIs themselves must comply with standards, but they must also have safeguards to prevent them from being tampered with in unforeseen ways.

NFV and 5G: NFV decouples software from hardware by virtualizing network functions such as firewalls, load balancers, and routers and operating them as software. This reduces the need for many expensive hardware components and can also shorten installation timeframes, allowing customers to receive revenue-generating services sooner.

By virtualizing appliances within the 5G network, NFV allows the 5G infrastructure. This includes network slicing, which allows numerous virtual networks to function simultaneously. Other 5G difficulties can be addressed with NFV by customizing virtualized computing storage, and network resources based on applications and consumer segments.

Security in Network Function virtualization (NFV)

The main idea behind virtualization is to decouple a system's service model from its physical realization to use logical instances of the physical hardware for different purposes.

Since the number of services or virtual functions is a growing concern as it is related to the manual configurations of the virtual systems or VNFs that can lead to potential security breaches due to the increased complexity with the growth of the systems.

Virtualization can highly increase user, service, and network security. A primary mechanism is to use slicing to separate the traffic of different services or network segments based on security priorities.

NETWORK FUNCTION VIRTUALIZATION (NFV)

Software is used to implement the functions of network elements or, more broadly, network services such as firewalls and gateways (SW). On the other hand, this software frequently operates on specialized and consequently proprietary hardware (HW), sometimes based on a proprietary operating system (OS).

The service is provided by numerous interacting NFs in the case of the CSCF. Individual services such as P-CSCF, etc. must be combined into one overall service by a central logic for this to happen. This is referred to as orchestration (instrumentation). This approach, which is still widely used to deploy network elements and network services on proprietary hardware, results in relatively high acquisition prices and a network architecture that is mainly set in function. To address these issues, the "Network Functions Virtualization (NFV)" concept was developed and standardized, particularly from the perspective of network operators. It is predicated on the premise that network functions are totally performed in software and so may be implemented on conventional hardware. As a result, tried-and-true IT virtualization techniques, including virtual machines (VM) and their coexistence on ordinary server hardware, can be used.

In 2012, the ETSI Industry Specification Group for NFV (ISG NFV) addressed this issue and defined NFV as follows: "Network Functions Virtualization aims to change the way network operators architect networks by advancing standard IT virtualization technology to consolidate many network equipment types onto industry standard high volume servers, switches, and storage, which could be located in data centers, network nodes, and end-user premises,... It entails the implementation of network services in software that can run on a variety of industry-standard server hardware and that can be transferred to or instantiated in numerous network locations as needed without the need for new equipment."

The usage of NFV can bring several benefits to network operators such as:

- Less expensive equipment

- New network capabilities and performance enhancements may be introduced more quickly now that only SW, not HW, is used

- Production, test, and reference environments all share the same HW infrastructure

- Excellent scalability

- Access to the market for software-only vendors

- Ability to modify network design in real-time to current traffic and its distribution within the Network

- Multiple network operators using the same HW

- Lessening the use of electricity

Due to the uniform HW platform, lower planning, provisioning, and operating expenses are possible.

- Using IT orchestration tools and reusing VMs, automate installation and operation.

  - The SW upgrading has been simplified.

- Creating synergy between network operations and information technology.

VNFs and NFs must be appropriately instantiated, monitored, and operated for NFV to work correctly (e.g., modulation, coding, multiple access, ciphering, etc.). In fact, the NFV framework comprises of software implementations of network functions (VNF), hardware (industry standard high volume servers) known as NFV Infrastructure (NFVI), and an architectural framework for virtualization management and orchestration. Some NFs may necessitate the use of hardware accelerators in order to meet real-time requirements. The accelerators take over computationally complex and time-sensitive activities that NFVI can't yet handle. As a result, not only can traffic be offloaded from NFVI, but latency requirements can also be met.

Note: In the context of virtualization, physical and logical pathways between endpoints (e.g. devices) in the Network must be differentiated.

The significant advantages of NFV are the lower capital and operational costs and the faster time to market. VNFs must, however, be portable between vendors and coexist with hardware-based network platforms in order to reap these benefits.

Multi-Access Edge Computing (MEC)

MEC is a key component in 5G design. MEC is a cloud computing evolution that moves applications from centralized data centers to the network edge, bringing them closer to end users and devices. This effectively eliminates the long network path that previously separated the user and the host in terms of content delivery.

This technology isn't unique to 5G but is critical to its effectiveness. Low latency, high bandwidth, and real-time access to RAN information are features of the MEC that separate the 5G design from its predecessors. Operators will need to use new ways of network testing and validation due to the convergence of the RAN and core networks.

5G networks that follow the 3GPP 5G specifications are suited for MEC implementation. The 5G specifications define the edge computing enablers that allow MEC and 5G to work together to transport data. The MEC architecture's distributed processing capability will better enable the enormous volume of connected devices inherent in 5G deployment and the advent of the IoT, in addition to the latency and bandwidth benefits.

MOBILE EDGE COMPUTING

Mobile edge computing (MEC) is the ability to host third-party applications at the mobile Network's edge, on edge hosts placed at radio nodes, aggregation points, or the core network's edge. For 5G operators and applications, MEC opens up new possibilities. For starters, it enables more significant support for low-latency apps by locating them close to their customers, minimizing the need for application traffic to cross the core network. Second, mobile operators can give mobile edge services to third-party mobile edge applications, such as location and radio information, to help them improve their performance and responsiveness. The following components of MEC's security characteristics can be seen.

- Mobile edge hosts outside the mobile operator's premises: Because mobile edge apps' data traffic does not pass through the core network, lawful interception and traffic accounting appear to be required at the mobile edge hosts. Suppose mobile edge hosts are placed at radio nodes or aggregation sites. In that case, lawful interception and traffic accounting will have to occur somewhere other than the mobile operator's premises, such as a stadium, shopping mall, college, hill, or rooftop. This raises the possibility of unlawful eavesdropping, fraudulent billing, and other attacks on the authorized interception and traffic accounting services. MEC's virtualization environment is even more challenging to safeguard physically than a typical cloud environment because of the multiple cloudlets (mobile edge hosts) located outside the operator's premises.

- Mobile edge services: The mobile operator can supply mobile edge services through MEC to third-party mobile edge applications. Mobile edge services must be able to authenticate mobile edge applications in order to ensure valid access to their services. As a result, credentials exposed outside the main Network may be at risk.

- Service continuity of mobile edge applications during UE handover: In the event of UE handover, the UE must be served by the nearest instance of the mobile edge application in order to retain low-latency communication and context awareness. As a result, either an application context transfer from the current instance of the mobile edge application to the next instance or a transfer of the mobile application instance itself from the current mobile edge host to the next one is required to ensure service continuity. In any instance, the source and destination must mutually authenticate each other, and the transfer must be secure.

- UE authentication and re-authentication: One of MEC's main promises is low-latency connectivity. In pre-5G mobile networks, the present UE authentication and re-authentication approach may need to be improved to reduce or eliminate the disturbance to low-latency communication, particularly during a UE handover.

DISTRIBUTED CLOUDS

A distributed cloud architecture employs several clouds to meet compliance standards, performance requirements, or edge computing while being controlled centrally by the public cloud provider.

A distributed cloud service is a public cloud that runs in various places, such as multiple data centers.

- The infrastructure of the public cloud provider
- In the data center of another cloud provider
- Hardware from a third party or a colocation center

Although there are different locations and countries involved, all cloud services are controlled as a single entity from a single control plane that manages the variations and inconsistencies that come with a hybrid, multi-cloud environment.

This service distribution allows an organization to fulfill very particular response time and performance needs, regulatory or governance compliance mandates, or other demands that require cloud infrastructure to be situated outside the cloud provider's regular availability zones.

The IoT and edge computing have fueled the expansion of distributed cloud deployments. Artificial intelligence (AI) applications that transmit huge amounts of data from edge locations to the cloud require cloud services to be as close to the edge as possible, and relocating cloud resources to the edge location can considerably improve performance.

Furthermore, an ever-increasing number of government rules, such as the EU's GDPR, may require data to be stored in certain jurisdictions that a given public cloud provider may or may not support, necessitating the use of a dispersed cloud.

Distributed clouds can benefit by putting cloud services closer to a specific user, application, or data.

- Decreased latency
- Congestion on the Network was reduced or eliminated
- QoS guarantees for mission-critical apps and mobile users

Benefits of Distributed Cloud

A distributed cloud architecture has numerous advantages. Following are some of the points that stand out:

- A rise in compliance. Workloads and data can be located where they need to be to meet regulatory requirements because they are distributed by nature.

- An increase in the amount of time that the system is available. Because cloud services can be isolated—even untethered from the main cloud—as needed, they can be isolated from a crashed system to provide redundancy.

- Scalability: Adding VMs or nodes as needed allows for rapid scalability and improves the cloud system's overall availability.

- Flexibility: Distributed clouds make installing, deploying, and debugging new services easier.

- Processing time is reduced. By combining the computing of numerous computers for a given task, distributed systems can be faster. Furthermore, the distributed cloud allows for more responsive messaging for specific geographic areas.

- Performance: As opposed to centralized computer network clusters, the distributed cloud can deliver superior performance and lower costs.

How Does a Distributed Cloud Work?

Services are "distributed" to specified locations in a distributed cloud to reduce latency, and these services have a single, consistent control point across public and private cloud environments. An organization can experience large performance gains by lowering latency and reducing the risk of an outage by eliminating latency concerns and reducing the total risk of outage or control plane inefficiencies.

A distributed cloud distributes not just an application but the full computing stack to the locations where it's needed, whether it's on-premises or in a public cloud provider, or in a third-party colocation facility. This distributed architecture appears to the consuming cloud customer as a single cloud entity. The cloud provider administers all aspects of the distributed cloud from a single control plane.

All cloud operations, including security, availability, upgrades, and governance of the entire distributed infrastructure, remain the responsibility of the public cloud provider. Distributed cloud, according to Gartner, fixes what hybrid cloud and multi-cloud fail to do.

What Are Use Cases for Distributed Clouds?

From smart edge computing to easing management of multi-cloud systems and hybrid deployments, distributed clouds have a wide range of applications. The following are examples of common use cases:

Everything from easier multi-cloud management to better scalability and development velocity to the deployment of cutting-edge automation and decision support applications and capabilities is aided by distributed cloud and edge computing.

- Edge/IoT. IoT is using AI and machine learning (ML) to improve automobile manufacture, analyze medical imaging, and smart buildings and smart cities that find the shortest route to parking and turn off the heating after the last employee has left for the evening, with new uses for video inference and facial recognition being developed on a daily basis. Many of these applications would be hampered if data had to be sent from the edge to the cloud or a data center for processing and analysis.

- Optimizing the content. Distributed clouds can successfully function as a content delivery network (CDN), enhancing streaming and lowering web page load time latency while providing the best possible user experience for a wide range of applications.

- Demand-based scaling Distributed clouds make it possible to expand to new areas without having to develop new infrastructure. As the organization's demands evolve, the cloud footprint can expand in lockstep to meet them.

- Management of a single pane of glass. Adopting a distributed cloud strategy improves visibility into a hybrid, multi-cloud deployment and the ability to manage all infrastructure as a single cloud from a single console using a single set of tools.

- Comply with legal requirements. Local, federal, and international data privacy laws might dictate where a user's personal information is held and whether it can travel outside of that country. When data cannot be moved to a public cloud provider, the data can be effectively transferred to the public cloud provider, ensuring that governance and legal criteria are followed, and that data is handled as efficiently as possible with the least amount of latency.

What Are The Challenges of the Distributed Cloud?

Managing an organization with a multi-site cloud deployment comes with its own set of issues, such as:

- Bandwidth. For each site in a widely dispersed multi-cloud system, there may be a variety of connectivity models. As a result, bringing more computation to the edge might place a strain on existing broadband connections, necessitating upgrades or adaptations to match higher throughput demands.

- Security. Because resources might be spread around the globe and collocated with existing business servers and storage resources, securing a distributed cloud creates new issues for both cloud providers and end-users.

- Data security. Backup and business continuity plans for dispersed data resources may necessitate a rethinking of backup and recovery procedures to ensure data remains in the intended locations.

The distributed micro cloud system is made up of several micro clouds that are deployed at the Network's edge. To prevent repeated redundant transmission of user-requested material in the Network, service content in the core cloud data center can be distributed and cached to the local micro cloud server in advance.

Distributed Cloud Infrastructure

This is a collection of various-sized cloud data centers located in global, national, local/regional, and maybe access locations, all of which are connected to the Network and managed by a central orchestration and management system. The infrastructure specifications on the various sites may vary depending on the use cases and applications onboarded. Furthermore, multiple infrastructure providers may be present on the same site.

The Role of Distributed Cloud Computing in 5G Networks

In the typical cloud computing concept, remote sites use the cloud as a hyper-scalable centralized virtual data center. The cloud computing paradigm is transformed in the opposite direction with 5G, transforming it into a dispersed cloud. A distributed cloud is a cloud execution environment geographically scattered across numerous sites, with requisite communication in between, is managed as a single entity, and applications view it as such.

Distributed Cloud Computing's Role in 5G Networks Remote sites use the cloud as a hyper-scalable centralized virtual data center in the typical cloud computing concept. The cloud computing paradigm is transformed in the opposite direction with 5G, transforming it into a dispersed cloud. A distributed cloud is a cloud execution environment that is geographically scattered across numerous sites, with requisite communication in between, is managed as a single entity, and applications view it as such.

SDN, NFV, and 3GPP technologies enable multi-access and multi-cloud capabilities in a distributed cloud consisting of a central and edge cloud. It can have several edges, and workload placement will be determined by where the application's requirements can be met for the least amount of money. The abstraction of cloud infrastructure resources, which hides the complexity of resource distribution from an application, is a critical feature of the distributed cloud.

CYBER THREAT INTELLIGENCE IN THE 5G ARCHITECTURE

Network operators can reduce the attack surface using Threat Intelligence with four key concepts in the 5G architecture. These are Multi-access Edge Compute, Data Network (DN), Network Slicing, and Virtualization.

1. MEC deployment

- Closer to the RAN and UEs to isolate and mitigate internal Botnets and DDoS attacks

- Separation from Core to isolate attacks

- Scales for UE Density

2. DN deployment

- Closer to Public Internet to isolate and mitigate external volumetric DDoS attacks and block malicious payloads.

- Separation of LAN services from Core

3. Network Slicing

- Optimizes ATI. Heuristics, anomaly detection, and ML are applied to traffic sets, reducing false positives.

4. Virtualization

- Service Chaining to security services can be applied per application and per enterprise

- Scales for Massive IoT

- Orchestration provides dynamic instantiation of VNFs to respond to security incidents.

- Resources can be dynamically allocated for surges in bandwidth, signaling rates, and UE density due to cyberattacks. Orchestration of ATI delivers faster Incident Response (IR).

CHAPTER 5

Differentiating 4G and 5G on a Security Basis

When comparing 5G to 4G, it's easy to see what 5G offers. Latency is (1 ms vs 10–50 ms), throughput is (20 vs 2 Gbps), spectral efficiency is (100 vs 30 bps/Hz), density is (1M vs 100K conns/km2), traffic capacity (1000 Mbps/m^2 vs 10 Mbps/m^2), and network energy efficiency are all improved over 4G. (15% savings).

According to one source, 5G promises a data throughput of up to 10 Gbps, 1-millisecond latency, 1000× bandwidth per unit area, and 100× device density over 4G, 99.999% availability, 100% coverage, 90% energy savings, and up to 10-year battery life for low-power Internet of Things (IoT) devices.

FROM 4G TO 5G

In comparison to 4G networks, which use IPsec tunnels between protocol components to ensure network domain security (NDs), 5G networks use transport layer security (TLS) and application layer security (AES) as well Service-Based Architecture (SBA). All network functions must support TLS with the client and server-side certificates. Within an operator domain, TLS will assure transport security. NDS can be implemented using NDS/IP. The operator decides whether or not to apply for cryptographic protection in NDS. It depends on whether component interfaces have been physically protected in trustworthy locations.

DOI: 10.1201/9781003264408-5

When the 5G core is built on SBA principles, 5G network services are planned to be modular, with interfaces built on SBA principles as well. According to this strategy, the Common Application Programming Interfaces (API) Framework (CAPIF) will secure all northbound (API), allowing for a single protection mechanism to be utilized for all system functions. In 4G, however, protection of protocol component northbound APIs is scattered.

ADVANTAGES OF 5G OVER 4G NETWORKS

In comparison to 4G networks, 5G networks will offer the following advantages:

- In addition to the 5G-Authentication Key Protocol (AKA) security methods, 5G introduces certifications for IoT devices.

- Integrity protection for user plane traffic is possible with 5G. In 4G, user plane security is disabled or enabled for all dedicated radio bearers. In contrast, in 5G, selective protection per PDU (Protocol Data Unit) session is possible. This feature will require additional resources at the gNB and the core network user plane function (UPF).

- If there is no security context in place in 4G, the device International Mobile Subscriber Identity (IMSI) is not secured. SUPI (Subscription Permanent Identifier), on the other hand, is safeguarded in 5G using asymmetric cryptography.

- When a device is in a visited network, the home control feature in 5G security allows the network to verify the device's location. This precaution was taken to avoid device location spoofing attacks. This 4G network vulnerability includes sending bogus signaling messages to seek the device's identity and location, then intercepting ongoing traffic.

- The notion of "unified authentication" is included in 5G security, which means that 3rd Generation Partnership Project (3GPP) (5G RAN) and non-3GPP access networks (e.g., WiFi) must use the same authentication mechanisms. The authentication mechanism for 3GPP access can give keys for establishing security in non-3GPP access that is not trusted. Regardless of the access network type, the User Equipment (UE) and networks will support both Extensible Authentication Protocol (EAP)-AKA and 5G AKA authentication. Only new WiFi equipment that supports the use of foreign keys from 3GPP networks will be able to do so.

- Against bidding down assaults, 5G offers a mitigation technique. It prevents a phony base station from convincing UEs that the base station lacks support for a specific security feature, forcing them to use an older mobile network technology. IMSI catchers are the name for such base stations. To simulate a base station in 5G, all coded and protected non-access stratum (NAS) messages would need to be acquired; thus, IMSI catchers will have difficulty succeeding.

- The 5G UE and the home network (HN) can decide which mobile technologies, such as 4G, EDGE, and 5G, can be used. After the device is turned off, the settings will be remembered and restored. Only emergency services can override this choice and use any technology supported by the available device.

PRIVACY AND INTEGRITY CIPHER

UE management in 5G is divided into two categories: NAS and access stratum (AS). The NAS layer protocol handles the connection between the UE and the core network (Access and Mobility Management Function [AMF]), whereas the AS layer protocol uses the Radio Resource Control (RRC) protocol to regulate the radio layer between the UE and the gNB. Security strives to provide RRC messages and IP packets securely, while NAS security ensures that signaling between UE and AMF may be delivered securely on the control plane.

To access the 5G network, the UE must undertake an initial connect procedure that initiates all protocol levels from the NAS layer to the AS layer. We presume that the UE and the gNB have already established a radio link in our approach. The initial attached procedure with NAS security and AS security goes like this:

1. To identify the supported security algorithms included in the Security Capabilities, the UE directs the Attach Request message (ARM) with the SUPI or SUCI.

2. Mutual authentication is established by the authentication and key agreement. When the AMF receives an authentication challenge from the UDM, it sends an Authentication Request message to the UE that includes a random nonce (RAND) and an authentication token (AUTN). The UE validates the authentication token, computes the network-verified response RES, and returns it.

3. The AMF picks ciphering and integrity algorithms based on the UE Security Capabilities and creates the NAS Security Mode Command message using the specified security algorithms. Moreover, it delivers the integrity-protected NAS Security Mode Command message to the UE to enable the NAS security mechanism. A NAS Security Mode Complete message appears on the UE. The NAS Security Mode Command message is protected integrity-wise but not ciphered, as you can see. It also displays the specified security algorithms and a replay of the original UE Security Capabilities to guard against algorithm downgrade attempts.

4. The network provides an IP address with the Attach Accept message. It initiates the AS security mechanism by including UE Security Capabilities in the Attach Accept message.

5. The gNB creates the AS Security Mode Command message based on the selected security algorithms in the Attach Accept message and transmits it to the UE to start the AS security mechanism. The UE sends an AS Security Mode Complete message. The AS Security Mode Complete message is integrity protected but not ciphered, which is something to consider.

As seen from the foregoing study, the choice of security algorithm is critical for the security protection of air interface signaling. The New radio Encryption Algorithm (NEA) and New radio Integrity Algorithm (NRIA) are two ciphering and integrity protection algorithms supported by 5G. (NIA). NEA1 and NIA1 use SNOW 3G, AES is used by NEA2 and NIA2, and ZUA is used by NEA3 and NIA3. The NEA0 and NIA0 null algorithms disable security, allowing data to be sent unencrypted. They allow emergency calls to be made even if a valid USIM and, as a result, a valid key is unavailable. Integrity protection is essential for ensuring the authenticity of sent communications and demonstrating that both parties have the right keys constantly.

AUTHENTICATION KEY AGREEMENT (AKA)

The AKA technique can be used to ensure secure communication in any cellular network. AKA stands for Authentication and Key Management, which entails mutual authentication between the user device and the network and the generation of crypto keys to safeguard data in the U-plane and C-plane. Each telco "G" specifies an authentication technique that allows only authorized users to connect to the network while rejecting unauthorized ones. For 4G LTE, 3GPP defined EPS-AKA, and for 5G, three authentication techniques were described.

1. 5G-AKA: 5G-Authentication and Key Management

2. EAP-AKA: Extensible Authentication Protocol-Authentication and Key Management

3. EAP-TLS: Extensible Authentication Protocol-Transport Layer Security

Need for New AKA Procedures in 5G

Any new technology's success depends on its ability to address security and privacy concerns. The availability of 5G networks is becoming a reality worldwide, and new use cases are emerging. Security experts and researchers have thoroughly investigated previous "G" RANs network security and privacy vulnerabilities. The following are a few examples of such problems.

Network spoofing: To lure UE away from its authentic cellular network and register with the counterfeit base station, a faked base station can broadcast a new tracking area code with greater signal strength.

Lack of confidentiality: Users' privacy may be jeopardized if specific OTA (Over The Air) signaling messages are intercepted. For example, unencrypted paging data can be utilized to detect the presence of a specific user and even trace them to a particular area.

To address these concerns, standardization organizations such as the 3GPP have specified an AKA protocol and processes that provide user authentication, signaling integrity, and signaling confidentiality, among other security features. The 3GPP AKA protocol is based on a challenge-and-response authentication mechanism that uses a symmetric key shared by the User and the Network. Crypto keys are obtained to safeguard subsequent communication between a User and a serving network (SN), including C-plane and U-plane data, after mutual authentication between a User and a HN.

LTE AUTHENTICATION PROCEDURE

From an authentication perspective, a cellular network (Figure 5.1) consists of three main components: UEs, a SN, and a HN.

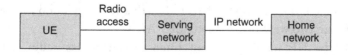

FIGURE 5.1   Cellular network architecture.

Each UE has a Universal Integrated Circuit Card (UICC) on which is installed at least one universal subscriber identity module (USIM) application, which maintains a cryptographic key that is shared with the user's HN. Radio access equipment, such as an Evolved NodeB (eNodeB) base station and Mobility Management Entities (MMEs), is part of a 4G providing network. Through radio interfaces, the UE communicates with an SN. Authentication servers, such as the home subscriber server (HSS), which stores user credentials and authenticates users, are typically found in a 4G HN. The essential entities that are connected through an IP network are collectively referred to as the Evolved Packet System (EPS); communication between SNs and an HN is based on IP.

4G EPS-AKA

After the UE finishes the RRC procedure with eNodeB and sends an Attach Request message to the MME, the EPS-AKA is activated. The MME directs an Authentication Request to the HSS on the HN, which includes the UE identity (i.e., IMSI) and the SN identifier. The HSS uses the shared secret key, Ki (shared with the UE), to generate one or more authentication vectors (AVs), which are subsequently transmitted back to the MME in an Authentication Response message. An AV contains an authentication (AUTH) token and an expected authentication response (XAUTH) token.

The MME directs an Authentication Request to the UE, including the AUTH token, after receiving an Authentication Response message from the HSS. By comparing the AUTH token to a created token based on Ki, the UE verifies it. If the validation is successful, the UE deems the network to be legitimate and sends an Authentication Response message to the MME, along with a response (RES) token produced using Ki.

The MME compares the RES token to a token representing an expected response (XRES). If they're equal, the MME executes key derivation and sends a Security Mode Command message to the UE, which then derives the keys needed to protect subsequent NAS signaling signals. The MME will also transmit the eNodeB a key that will be used to generate the keys for protecting the RRC channel. Following the UE's derivation of the relevant keys, communication between the UE and the eNodeB is protected.

There are two weaknesses in 4G EPS-AKA.

First, the UE identity is sent unencrypted across radio networks. Although a temporary identifier (e.g., Globally Unique Temporary Identity,

GUTI) can be used to conceal a subscriber's long-term identity, researchers have discovered that GUTI allocation is flawed. GUTIs are not changed as frequently as they should be, and GUTI allocation is predictable (e.g., with fixed bytes). When responding to an Identity Request message from a network, the UE's permanent identity can be delivered in clear text in an Identity Response message.

Second, when an SN consults an HN during UE authentication, the home network supplies AVs, but it is not a factor in the authentication decision. The SN is the only one who can make such a judgment.

5G AKA AUTHENTICATION PROCEDURE

To make 5G authentication both open (e.g., with EAP support) and access-network agnostic (e.g., supporting both 3GPP and non-3GPP access networks such as WiFi and cable networks), a unified authentication framework has been designed (see Figure 5.2).

Note: When EAP is utilized (for example, EAP-AKA' or EAP-TLS), EAP authentication is done via the Security Anchor Function (SEAF) between the UE (an EAP peer) and the AUSF (an EAP server) (functioning as an EAP pass-through authenticator).

When authentication is performed across untrusted, non-3GPP access networks, a new entity called the Non-3GPP Interworking Function (N3IWF) is required to act as a VPN server, allowing the UE to connect to the 5G core via IPsec (IP Security) tunnels.

Several security contexts can be established with a single authentication execution, allowing the UE to migrate from a 3GPP access network to a non-3GPP network without reauthenticating.

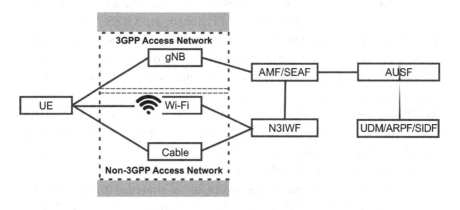

FIGURE 5.2 5G Authentication framework.

5G AKA

New authentication-related services are defined by 5G. The AUSF, for example, provides authentication using Nausf UEAuthentication, and UDM provides authentication through Nudm_UEAuthentication. In procedure generic messages such as Authentication Request and Authentication Response are used instead of the actual authentication service names for convenience. In addition, an authentication vector contains a collection of data.

After receiving any signaling transmission from the UE in 5G-AKA, the SEAF may begin the authentication operation. If the SN has not assigned a 5G-GUTI to the UE, the UE shall give the SEAF a temporary identification (a 5G-GUTI) or a permanent encrypted identifier (a SUCI). The SUCI is an encrypted version of the SUPI that uses the HN's public key. As a result, the permanent identity of a UE, such as the IMSI, is never broadcast in plain text over 5G radio networks. This capability is seen as a significant security upgrade over previous generations, such as 4G.

The SEAF initiates authentication by sending a request to the AUSF, which checks that the providing network seeking the authentication service is authorized first. The AUSF sends an authentication request to UDM (Unified Data Management)/ARPF (Authentication Credential Repository and Processing Function) if it is successful. If the AUSF provides a SUCI, the SIDF will be called to decrypt the SUCI and retrieve the SUPI, which is then used to pick the authentication method for the subscriber. It is 5G-AKA that has been chosen and will be executed in this scenario.

UDM/ARPF initiates 5G-AKA by sending an authentication vector to the AUSF, which includes an AUTH token, an XRES token, the key KAUSF, and the SUPI, if applicable (for example, when a SUCI is provided in the associated authentication request), among other data.

The AUSF generates a hash of the expected response token (HXRES), saves the KAUSF, and sends the authentication response, along with the AUTH token and the HXRES, to the SEAF. In this authentication response, the SUPI is not sent to the SEAF. After successful UE authentication, it is solely forwarded to the SEAF.

The SEAF stores the HXRES, and the AUTH token is sent to the UE in an authentication request. The secret key is shared with the HN, and the UE validates the AUTH token. The UE considers the network to be authenticated if validation passes. The UE completes the authentication by computing and delivering a RES token to the SEAF, which the SEAF validates. Following the success, the SEAF sends the RES token to the AUSF for validation. It's worth noting that the AUSF, which is part of an HN, makes the final authentication decision. The AUSF computes an anchor key (KSEAF) and sends it to the SEAF, along with the SUPI, if appropriate, if the UE's RES token is valid. The AUSF also notifies UDM/ARPF of the authentication results so that they can record the occurrences, for example, for auditing purposes.

The SEAF derives the AMF key (KAMF) from the KSEAF (and subsequently deletes the KSEAF) and delivers the KAMF to the co-located AMF. The AMF will then derive (a) the confidentiality and integrity keys needed to protect signaling messages between the UE and the AMF from the KAMF and (b) another key, KgNB, which is sent to the Next Generation NodeB (gNB) base station for deriving the keys needed to protect subsequent communication between the UE and the gNB from the KAMF. The long-term key, the root of the key derivation hierarchy, is held by the UE. As a result, the UE can deduce all of the above keys, resulting in a shared set of keys between the UE and the network.

The following are the main differences between 5G-AKA and 4G EPS-AKA:

Because of the new service-based architecture in 5G, the authentication entities are different. The SIDF, in particular, is novel; it does not exist in 4G.

The UE always encrypts it with the HN's public key before sending the UE's permanent identification to a 5G network. In 4G, the UE always broadcasts its permanent identity to the network in clear text, allowing it to be stolen via the radio connections by either a malicious network (e.g., a false base station) or a passive adversary (if communication over radio links is not protected).

In 5G, the HN makes the final decision on UE authentication (e.g., the AUSF). Additionally, the results of UE authentication are provided to UDM for logging. During 4G authentication, an HN is just consulted to create authentication vectors; it does not make decisions based on the authentication findings.

Because 5G introduces two intermediate keys, KAUSF and KAMF, the key hierarchy is longer in 5G than in 4G.

Note: In 5G, KSEAF is the anchor key, whereas, in 4G, KASME is the anchor key.

EAP-AKA

Another authentication mechanism available in 5G is EAP-AKA. It's also a challenge-and-response system that relies on a shared cryptographic key between a UE and its HN. It has the same level of security as 5G-AKA, such as mutual authentication between the user and the network. Its communication flows differ from those of 5G-AKA because it is based on EAP. EAP communications are contained in NAS messages sent between the UE and the SEAF and in 5G service messages sent between the SEAF and the AUSF. The following are some of the other distinctions between 5G-AKA and EAP-AKA.

The SEAF plays a slightly different role in the authentication. EAP messages are exchanged between the UE and the AUSF through the SEAF in EAP-AKA, which transparently forward the EAP messages without being involved in any authentication decisions. The SEAF in 5G-AKA also verifies the UE's authentication response and may take action if the verification fails.

The key derivation is slightly different. The KAUSF is computed by UDM/ARPF and communicated to the AUSF in 5G-AKA. The AUSF generates the KAUSF partly from the keying materials obtained from UDM/ARPF in EAP-AKA. The AUSF generates an Extended Master Session Key (EMSK) using the keying materials received from UDM in accordance with EAP, then uses the first 256 bits of the EMSK as the KAUSF.

EAP-TLS

EAP-TLS is a subscriber authentication protocol described in 5G for limited use cases such as private networks and IoT contexts. EAP-TLS is done between the UE and the AUSF through the SEAF, which acts as a transparent EAP authenticator by sending EAP-TLS messages back and forth between the UE and the AUSF when UDM/ARPF selects it as the authentication mechanism. Both the UE and the AUSF can validate each other's certificate or pre-shared key (PSK) if it was established in a previous TLS handshaking or out of band to achieve mutual authentication. An EMSK is derived at the end of EAP-TLS, and the first 256 bits of the EMSK are used as the KAUSF. The KAUSF is used to generate the KSEAF, which is then used to derive additional keying materials needed to safeguard communication between the UE and the network, the same as in 5G-AKA and EAP-AKA.

EAP-TLS is significantly different from 5G-AKA and EAP-AKA. It uses a different trust paradigm to create trust between a UE and the network. A mutual authentication between a UE and a 5G network is achieved mainly through mutual trust of their public key certificates in EAP-TLS, noting that TLS with a PSK is possible but rarely utilized except for session restart. Such confidence is entirely predicated on a symmetric key shared between a UE and the network in AKA-based approaches.

Such a fundamental distinction is significant because EAP-TLS eliminates the need to keep a huge number of long-term keys in the available HN (e.g., in UDM), lowering operational risks in the symmetric key management life cycle. EAP-TLS, on the other hand, adds new overhead to certificate administration, such as issuance and revocation.

We can compare 4G and 5G authentication methods, highlighting the differences between the two technologies. For example, 5G authentication has dissimilar entities from 4G because 5G implements service-based architecture. Another major differences include the trust models in methods based on EAP-TLS or AKA protocol, the entities that make authentication decisions, and the anchor key hierarchy

SEAF or Anchor Key

With the new SEAF in 5G, the concept of an anchor key is introduced.

When a device moves between multiple access networks or even SNs, the SEAF allows it to re-authenticate itself without having to go through the complete authentication process (for example, AKA authentication). Various mobility services decrease the signaling burden on the HN HSS.

The SEAF and AMF could be split up or combined. The SEAF capability is co-located with the AMF in 3GPP Release 15.

The Granularity of the Anchor Key Binding to the Serving Network

The anchor key KSEAF must be bound to the SN using the primary authentication and key agreement processes. The SN binding prohibits one SN from claiming to be another and hence offers implicit SN authentication to the UE.

This implicit SN authentication must be provided to the UE regardless of access network type, which means it applies to both 3GPP and non-3GPP access networks.

The anchor key binding will be accomplished by adding a "serving network name" parameter to the chain of key derivations that runs from the long-term subscriber key to the anchor key.

Home Control

Remember, The home control is not available in LTE.

In 5G, The home control is utilized to authenticate the device's position when the device is roaming. When the HN receives a request from a visiting network, it can verify that the device is truly in the network.

This was included to address vulnerabilities discovered in 3G and 4G networks, where networks might be spoofing and sending bogus signaling messages to the HN to obtain a device's IMSI and location. This data might then be used to listen in on phone calls and read text messages.

Network Exposure Function (NEF)

Beyond providing basic connectivity for data transit – the traditional "dumb pipe" – the next generation of telecommunications networks offers various services and capabilities. These include data on the location and reachability of connected devices, the ability to apply precise quality-of-service and billing policies to specific data flows, awareness of network load/congestion circumstances, and so on. These features are available in mobile core networks via an "exposure function."

The 5G NEF allows developers to access exposed network services and capabilities in a safe, reliable, and developer-friendly manner. The network domain provides access to both internal (i.e., within the network operator's trust domain) and external applications via a set of northbound RESTful (or web-style) APIs. The NEF is a function that works similarly to the 4G Service Capabilities Exposure Function (SCEF). To disguise the specific network technology from applications and user devices that can switch between 4G and 5G, a hybrid SCEF+NEF node is required. As a result, flexibility on the southbound network interfaces is a must for integration with 4G core endpoints via the Diameter protocol and with 5G core endpoints via the service-based interfaces defined in the new 5G Core SBA.

Authorized third-party developers and companies can use the accessible northbound APIs to establish their own network services on-demand. Application server interactions with policy and charging controls, network analytics, edge computing components, and network slicing may improve service assurance and network automation. The NEF can be used to create a multi-layer policy framework that allows policy decisions to be made at the application, business, and infrastructure levels.

CSPs, enterprises, device makers, and other participants must work together to capture the business value that may be fragmented across many use cases and services. In some circumstances, CSPs can generate total value by enabling end-user interactions. As a result, service providers become trusted partners' access platforms and enablers. CSPs can use network exposure capabilities to help their industry partners innovate and deliver value-added services, allowing them to establish trust. These cutting-edge technologies can benefit their users, such as higher productivity, lower costs, and automated business operations. This, in turn, can help drive more product and service innovation and create the potential for new business models that revolutionize companies across various industries.

The NEF allows network functions to be exposed to the outside world. External exposure includes monitoring capabilities, provisioning capability, policy/charging capability, network status reporting capability, and analytics reporting capability. The Monitoring feature keeps track of specific events for UE in 5GS and makes that information available to the outside world via the NEF. The Provisioning functionality allows an external party to provide information that a UE may use in 5GS. The Policy/Pricing feature is used to handle QoS and charging policies for UEs in response to external requests. The Analytics feature allows a third party to obtain analytics data provided by the 5G system.

NEF Features

The NEF is a key ingredient of 5GC SBA which adapts and transforms telecom protocols like RESTful APIs. Some exposure capabilities of NEF are highlighted below:

- 3GPP R15 (and beyond) NEF capability exposure.

  - Monitoring (e.g., AMF and UDM events).

  - Provisioning (e.g., UE data)

  - Policy and Charging

  - Core network internal capabilities for analytics.

- Advanced capability exposure

  - Edge computing

- IoT management

- On-demand end-to-end (RAN and Core) slice management.

- Call Management/TAS, and other APIs, e.g. Digital Assistant.

- API composition and orchestration

- API composition and orchestration

NEF has some important elements. It is worth mentioning the API Gateway and the API portal

Monitoring: Allows an external entity to subscribe to or request UE-related events of interest. The roaming state of a UE, UE loss of connectivity, UE reachability, and location-related events are all monitored events (e.g., location of a specific UE or identification of UEs within a geographical area). The AMF and the UDM are the two main entities providing access to this event data.

Provisioning allows an external entity to provide the 5G system with anticipated UE behavior, such as predicted UE movement or communication characteristics.

Policy and Charging: Manages QoS and charging guidelines for UE-based requests from third parties and sponsored data services. Although most NFs are active in some way in supporting the PCC Framework, the PCF is the most crucial entity in terms of Policy and Charging Control (PCC).

API Gateway Is an Application Programming Interface (API)
The API gateway is a crucial component of NEF. Its primary function is to enforce regulations and access controls between external parties and APIs. As a result of NEF's arrival, all requests would be routed through an API gateway to a specific service.

Portal for APIs
The API Portal serves as a resource for both providers and third parties. The portal explains which APIs are accessible for use, listing all of them and offering descriptions for each method.

The portal documentation should also include the authentication and authorization mechanism and use cases that highlight the business context and real-world implementations.

The portal also includes information on API lifecycles, eligibility to be an API consumer, pricing, API health insights (real-time monitoring), and more.

Using NEF for Event Exposure

The Monitoring Events feature is designed to keep track of certain events in the 3GPP system and to communicate that information to the NEF. It consists of mechanisms that allow NFs in 5GS to configure specific events, detect events, and report events to the requested party.

A roaming arrangement between the HPLMN and the VPLMN is required to allow monitoring functionalities in roaming scenarios. If the AMF/SMF in the VPLMN determines that an event report requires normalization, the AMF/SMF normalizes the report before delivering it to the NEF.

In 5GS, the set of capabilities required for monitoring will be available to NFs via NEF. Monitoring Events via UDM, AMF, SMF, NSACF, and GMLC allows NEF to configure a specific Monitor Event at UDM, AMF, SMF, NSACF, or GMLC, as well as report the event via UDM and/or AMF, SMF, NSACF, or GMLC. The AMF, GMLC, NSACF, or UDM is aware of the monitoring event or information and reports it via the NEF, depending on the specific monitoring event or information.

The following organizes the set of functional blocks into three groups. The first group runs in the Control Plane (CP) and has a counterpart in the EPC.

AMF (Core Access and Mobility Management Function): Connection and reachability management, mobility management, access authentication and authorization, and location services are all responsibilities of this position. It manages the EPC's MME's mobility-related elements.

SMF (Session Management Function): Controls all aspects of each UE session, including IP address assignment, UP function selection, QoS control, and UP routing control. Roughly corresponds to a portion of the EPC's MME and the EPC's PGW's control-related features.

PCF (Policy Control Function): Supports a unified policy framework to govern network behavior. Provides policy rules to Control Plane function(s) to enforce them. Accesses subscription information relevant to Unified Data Repository (UDR) policy decisions.

UDM (Unified Data Management): Manages user identity, including generation of authentication credentials. It contains part of the functionality in the HSS of the EPC.

AUSF (Authentication Server Function): It is essentially an authentication server. It contains part of the functionality in the EPC's HSS.

The second group also runs in the Control Plane (CP), but has no direct matching part in the EPC:

SDSF (Structured Data Storage Network Function): It is a "helper" service which stores structured data. It can be implemented by an SQL Database, in a microservices-based system.

UDSF (Unstructured Data Storage Network Function): It is a helper service used to store unstructured data. Could be implemented by a section of in a microservices-based system.

NEF (Network Exposure Function): It is meant to expose select capabilities to third-party services, including translation between internal and external representations for data. It can be implemented by an approach to in a microservices-based system.

NRF (NF Repository Function): A means to discover available services. It could be implemented by a "Discovery Service" in a microservices-based system.

NSSF (Network Slicing Selector Function): A means for selecting a network slice to serve a given UE. Network slices are essentially a way of partitioning network resources to differentiate services provided to different users. This is a crucial feature of 5G, which we will discuss in detail later.

The third group contains only one component that works in the User Plane (UP):

User Plane Function (UPF): Routes traffic between the RAN and the Internet according to the S/PGW combination in EPC. Besides packet forwarding, it is responsible for policy enforcement, lawful interception, traffic usage reporting, and QoS monitoring.

NEF FUNCTIONALITY

Independent functionality is supported via the NEF:

Capabilities and events exposure: NEF can securely expose NF capabilities and events to 3rd parties, Application Functions, and Edge Computing. NEF uses a standardized interface (Nudr) to the

Unified Data Repository to store and retrieve structured data (UDR) information.

Secure data transmission from an external application to the 3GPP network: It gives Application Functions a secure way to provide information to the 3GPP network, such as expected UE behavior, 5G-VN group information, time synchronization service information, and service-specific information. In that instance, the NEF may verify and approve the Application Functions and assist in throttling them.

Internal-external information translation: It converts data sent to and received from the AF and data sent to and received from the internal network function. It can, for example, translate between an AF-Service-Identifier and internal 5G Core data like DNN, S-NSSAI, etc.

NEF, in particular, handles network and user-sensitive information masking external AF's per network regulations.

- Redirecting the AF to a better NEF/L-NEF, such as when serving an AF request for local information exposure and recognizing a better NEF instance to satisfy the AF's request.

- The NEF receives information from various network functions (based on exposed capabilities of other network functions). NEF uses a defined interface to a Unified Data Repository to store the received data as structured data (UDR). The NEF can access and "re-expose" the stored data to other network functions and Application Functions and use it for other purposes like analytics.

- A NEF may additionally have a PFD Function: The PFD Function in the NEF may store and retrieve PFD(s) in the UDR, and shall provide PFD(s) to the SMF on SMF request (pull mode) or PFD management request from NEF (push mode).

- A NEF may support a 5G-VN Group Management Function: The 5G-VN Group Management Function in the NEF may store 5G-VN group information in the UDR via UDM.

- **Analytical exposure:** NEF may safely expose NWDAF analytics to third parties.

- **NWDAF data retrieval from a third party:** Data submitted by a third party may be gathered by NWDAF via NEF for the purpose of analytics production. NWDAF and AF use NEF to handle and forward requests and notifications.

- **Non-IP Data Delivery Support:** By exposing the NIDD APIs, NEF enables control of NIDD setup and delivery of MO/MT unstructured data.

A particular NEF instance may support one or more of the features mentioned above, and as a result, a single NEF may only support a subset of the APIs provided for capability exposure.

Note: The NEF can access the UDR located in the same PLMN as the NEF.

Support for Edge Computing

Edge computing allows the operator and third-party services to be hosted close to the UE's access point of attachment, resulting in more efficient service delivery due to lower end-to-end latency and network load.

Note: The edge computing typically applies to non-roaming and LBO roaming scenarios.

The 5G Core Network chooses a UPF near the UE. It sends traffic to the local Data Network through an N6 interface based on the traffic steering rules provided to the UPF. It could be found on the UE's subscription data, UE location, application function (AF) information, EAS information reported by EASDF, policy, or other associated traffic regulations.

Service or session continuity may be necessary due to the user or Application Function mobility, depending on the needs of the service or the 5G network.

An Edge Computing Application Function may be able to access network information and capabilities via the 5G Core Network.

Note: Depending on the operator deployment, some application functions may be authorized to interface directly with the Control Plane Network Functions (CPNF) with which they must interact. In contrast, others must use the NEF to expose themselves to the outside world.

Single or a combination of the following enablers can facilitate edge computing:

- Re-selection of the user plane (the 5G Core Network re-selects UPF to route user traffic to the local Data Network)

- Local Routing and Traffic Steering (the 5G Core Network chooses the traffic that will be routed to the local Data Network's applications)

- This includes using a single PDU Session with multiple PDU Session Anchor(s) (UL CL/IP v6 multi-homing) and using a PDU Session with Distributed Anchor Point (SSC mode 2/3)

- UE and application mobility is enabled through session and service continuity

- Using PCF or NEF, an Application Function can impact UPF (re) selection and traffic routing

- Network capability exposure: The 5G Core Network and Application Function will exchange data via NEF, directly or from the UPF

- QoS and Charging: PCF issues rules for QoS Control and Charging traffic routed to the local Data Network

- Support for the Local Area Data Network (LADN): The 5G Core Network allows users to connect to the LADN in a specific area where apps are installed

- Edge Application Server discovery and re-discovery

- Support for Edge Relocation, as well as the case of AF modification.

- Support for DNAI-based (I-)SMF (re)selection

IoT Management

To enable secure and fast access to IoT control and management of APIs in a single, robust, dedicated API gateway, a 5G NEF is required. API response times are critical for IoT use cases that require Ultra-Reliable Low Latency Communication (URLLC), such as robotic assembly. As a result, the 5G NEF platform must deliver exceptional performance.

SCEF is integrated with its 5G NEF capabilities (in one of the largest telco solutions) to provide operators with a seamless upgrade path, allowing them to easily move from 4G API exposure to 5G API exposure.

The power of network slicing, End-to-end network slicing is a crucial use case for operators looking to maximize and monetize their 5G investments.

A network slice can quickly meet a robot production line's availability and latency requirements. The dedicated network slice attributes must be configured to meet the production line's massive IoT traffic and low latency requirements. For example, massive UPF and Session Management Function (SMF) resources must be allocated near the endpoints at the operator's telco cloud's edge. The slice-related AMF can also be implemented in the central telco cloud because the manufacturing plant location is fixed from the 5G mobile network perspective.

The slice has a special NEF that only accepts API requests from the line's IoT control software; any other inquiries are refused. The slice-specific instantiation of NEF additionally contains its own API Gateway function to access the slice-specific APIs and preferred placement in the edge to reduce latency. APIs collect data from the control software and judge how to interact with the production line. The parameters and data in API queries are encoded using the http/2 protocol.

Similarly, because http/2 is extensively used for RESTFul (Representational State Transfer) APIs, the API answers to control software are in this protocol.

Several different use case apps may send many queries to the 5G core network's master-NEF. While these may impair the overall network, the slice-specific NEF will be unaffected and will continue to execute at a high level. The slice-specific NEF would continue to work even if the master NEF became overloaded.

This improves the service's security by preventing other applications from accessing the slice-related APIs. Similarly, API response time is improved because the gateway is dedicated to its specific use case.

Slice-specific network exposure will be a significant skill that operators may employ to generate new income and support many of the most recent IoT use cases.

CHAPTER 6

5G, IoT, and Cyber Risk

5G is distinguished by higher data rates (more than 10 gigabits per second), more capacity, and extremely low latency.

The architecture of 5G technology can be divided into three tiers.

1. Macro and microcells with appropriate base stations or hotspots to ensure ubiquitous connectivity among end devices or end-users.

2. The core network, which consists of routers, gateways, and other components, is responsible for gathering and relaying the data collected by the base stations.

3. The final link to the Internet could be via servers, data centers, or cloud infrastructures.

For various reasons, the combination of the Internet of Things (IoT) with 5G is a game-changer. The growth of IoT devices, paired with the deployment of 5G networks, opens up a slew of new possibilities for edge technology.

5G's ultra-fast connections and low latency are critical for intelligent automation in the digital future. This confluence will help emerging technologies like artificial intelligence (AI), driverless cars, virtual reality, and blockchain.

This junction, however, raises new security concerns for 5G and IoT technology. Although IoT security has improved over time, most connected equipment and gadgets were created with security as a secondary consideration, with minimal minimum criteria considered "good enough."

DOI: 10.1201/9781003264408-6

Massive scale is projected on eSIM (embedded-SIM) systems on the backs of trust-based networks and applications that authenticate without hardware security, which is truly remarkable. Given the growing importance of data security in today's environment, adding extra device and network-level security layers is vital, particularly for mission-critical applications like government, defense, and healthcare.

The number of linked things and smart gadgets in the IoT is continuously expanding in this digital age. In reality, 5G enables the present challenges occurring as a result of the increased number of network devices in terms of network response times and network resource management.

IoT

5G will fundamentally alter the way our global networks operate. It won't be long before global civilization—spanning industries, markets, and regions—must adjust to the new technology way of life. This new technology standard promises more than just improvements to current mobile communication systems.

Speed   Signal   Big Data   Internet   Network   Technology   Internet of things   Traffic

FIGURE 6.1   5G IoT.

The new network's data speed should be up to 20 gigabits per second, allowing for faster reaction times. The first cell phone with 1G network access costs 8 million times less than a 5G network to give you a sense of scale. It will also be feasible to send data in real-time with 5G. This means that 100 billion mobile devices will be connected at the exact moment around the world.

5G and IoT Security Concerns

IoT devices are becoming more accessible to individuals and businesses as 5G technology becomes more widely available. To enable IoT devices, several companies intend to deploy 5G networks. IoT devices can now run at higher speeds, with lower latency and lower costs, thanks to 5G. Using 5G without a private network or proper security measures, on the other hand, could jeopardize the privacy of businesses and individuals.

Suppose hackers acquire access to the 5G network. In that case, they might gain instant connectivity to every IoT device on the network, allowing them to access private data and use the devices as a Distributed Denial of Service (DDoS) attack tool. Because of the increased connectivity, accessibility, and coverage, this could lead to more IoT-based assaults. In addition, visibility into the 5G IoT network is limited, making it harder for cybersecurity teams to detect and remove malware before it infiltrates the network of linked devices.

Security and Privacy

The frequency spectrum used by 5G is comparable to that used by other remote-sensing devices. As a result, it could cause problems for weather forecasters and commercial aviation by interfering with satellites. 5G has the potential to be utilized for surveillance. Some governments have accused Chinese vendors of spying on their 5G gadgets on international users. In circumstances of mass spying, snoopers can also utilize 5G to capture metadata, and 5G has been criticized for being under-tested and claiming to disclose a large quantity of data. It's also projected that the amount of Denial of Service (DoS) and other assaults will rise. Due to 5G's high speed and latency, more operations will rely on Internet connections. IoT devices will grow increasingly common, and data will flow in the open and be stored on cloud servers, making them even more vulnerable.

Fog Computing

Because millions of devices are connected to the Internet, information security is a primary concern for organizations, governments, and individuals in today's world. Beyond stealing intelligence or interrupting company operations, hackers may readily uncover new weaknesses to exploit. Moreover, increasingly complex attacks make it significantly more difficult for systems to identify, protect, and respond to these threats.

IoT 5G has emerged as a solution for reducing such dangers, thanks to fog computing and fog computing architecture. Fog's distributed design protects linked systems from cloud to device, adding an extra layer of security by bringing computation, storage, networking, and communications closer to the services and data sources they serve and protect. Security is embedded in local surroundings rather than being a remote function.

Note: Fog computing is a computing architecture (Figure 6.2) in which a series of nodes receive data from IoT devices in real-time. These nodes process the data they receive in real-time, with millisecond reaction times. The nodes communicate analytical summary data to the cloud on a regular basis. Fog networking, often known as "fogging," is a word used by Cisco in 2014 to describe a decentralized computing architecture that functions as a cloud computing extension. Data storage and computation are dispersed between the cloud and the data source most logically and efficiently possible.

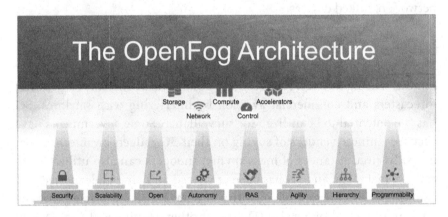

FIGURE 6.2   OpenFog architecture.

Cloud-based IoT and fog-based services are protected by fog nodes, which execute a wide range of security functions on any number of networked devices. Fog computing provides a secure distributed platform and execution environment for applications and services, manages and updates security credentials and malware detections, and provides timely software patches quickly and scalably.

CYBERSECURITY ISSUES

Many cybersecurity issues affect 5G, but we will focus on the three most severe ones in this section.

Latency

Many Machine-to-Machine (MTC), IoT, and Vehicle-to-Everything (V2X) communications will need to take place in situations with very low latency (less than one millisecond).

Furthermore, these settings must be exceedingly dependable and available. However, many Internet-based threats target access nodes such as low-powered access nodes and Long Term Evolution (LTE) nodes, making present 5G networks susceptible.

A DoS attack is able to render a system utterly inoperable. However, when sub-millisecond latency is required, the threshold for a successful attack is dropped. This gives a wide range of bad actors with few resources the potential to launch network-devastating attacks.

Note: A DoS attack aims to bring down a computer system or network so that its intended users are unable to access it. DoS attacks achieve this by transmitting information that causes a crash or by saturating the target with traffic.

Passive Eavesdropping

The number of devices connected to a network is rapidly expanding. Furthermore, each network contains a wide variety of devices.

As a result, network operators choose open-access auxiliary networks such as Wireless Local Area Networks (WLANs) because of their capacity requirements. However, the data transmitted within these WLANs are vulnerable to eavesdropping attempts and unauthorized access by users.

Taking advantage of the simplicity of low-power access points is one-way hackers can eavesdrop. This is the case because such access points lack the rapid and reliable authentication handover procedures required for 5G networks.

An attacker can monitor a specific user across networks, which extends this attack. It happens because the user's device transmits its International Mobile Subscriber Identity (IMSI) over the air and in an unencrypted form during the first connection phase on existing 5G networks. An attacker can then listen for these IMSIs in the network.

User Impersonation

Although this is a kind of eavesdropping attack, it has more potent consequences. The attacker just absorbs information intended for another user in passive eavesdropping.

On the other hand, the attacker can impersonate the user and send malicious messages or do malicious acts. This gives the attacker much more power over the situation because they can use their forged credentials to launch more attacks. The attacker's impersonation could endure for a long time, depending on the length of the access keys.

SOLUTIONS

Since we've covered some of the most critical concerns around 5G cyber-security, let's look at some feasible and practical solutions.

Network Slicing

This is referred to as virtualization when numerous virtual independent logical networks are created over the same physical infrastructure. Because this technology allows 5G operators to manage and configure each network slice individually, it should be employed. The physical network operator may occasionally lease these slices to different Mobile Virtual Network Operators (MVNOs).

This gives MVNOs even more flexibility because they can cater directly to their client's needs thanks to their specialized knowledge. This is beneficial because various entities have different needs despite sharing the same physical infrastructure. These requirements contain, but are not restricted to bandwidth, latency, privacy, QoS, and so on.

5G and network slice operators can utilize technologies like Software Defined Networks (SDN) and Network Function Virtualization (NFV) to enable network programmability in a real-world deployment. As a result, physical networks have the freedom to build several virtual networks that share the same physical infrastructure.

Notes: SDN is an approach to networking that uses software-based controllers or application programming interfaces (APIs) to communicate with underlying hardware infrastructure and direct traffic on a network. Network function virtualization, or NFV, is a method for network operators to save money and speed up service rollout by detaching functions like firewalls and encryption from dedicated hardware and shifting them to virtual servers.

Physical Layer Security

The basic idea here is to exploit the inherent randomness of the noise in the communication medium to limit the amount of information an attacker can decode.

Many examples of Physical Layer Security implementations, such as Orthogonal Frequency Division Multiplexing (OFDM), have been proposed. However, we will not cover it here a this is a highly technical topic.

The most important takeaway is that it can be used to protect network users from eavesdropping attacks.

Temporary Identification

Instead of providing their IMSI (which, as previously noted, could permit tracking), users disclose a temporary identity to non-trusted networks while roaming between trusted and untrusted networks.

The temporary identity could be generated on the trusted network and then shared by the user's device while inside the untrusted network, for example.

In today's 5G networks, methods based on the Extensible Authentication Protocol (EAP) are used. We will not discuss the EAP because it is too technical and beyond our scope.

CHAPTER 7

5G Security Risk

Because 5G uses network slicing, cloud-based resources, virtualization, and other upcoming technologies, 5G's security architecture provides for considerable performance gains and a wide range of applications.

However, new security risks and attack surfaces are introduced due to these additions and changes.

THE ADDED SECURITY THREATS

Increased attack surface: Because 5G connects millions, if not billions, more devices, it allows for more significant and more deadly attacks. The present Internet infrastructure's current and future vulnerabilities are merely amplified. With 5G, the potential for more advanced botnets, privacy violations, and speedier data extraction could increase.

More (Internet of Things) IoT means more issues: IoT devices are inherently unsafe, and security isn't always built-in. An attacker might exploit each insecure IoT device on a company's network is another potential security flaw.

Reduced network visibility: As 5G networks spread, they will become more accessible to mobile consumers and devices. This means there will be a lot more network traffic to deal with. Companies may not be able to get the network traffic visibility required to spot irregularities or assaults without a comprehensive wide area network (WAN) security solution in place, such as Secure Access Service Edge (SASE).

DOI: 10.1201/9781003264408-7

Increased supply chain and software vulnerabilities: 5G supply chains are currently and in the foreseeable future constrained. Vulnerabilities exist, which increases the risk of faulty and insecure components, especially as products are hurried to market. 5G is considerably more software-dependent than existing mobile networks, increasing the danger of network infrastructure exploitation.

5G RESILIENCE AND RISK

Quad countries' actions show that various technical, organizational, and policy steps are required to minimize risks associated with 5G technology, network implementation, and global supply chain security. A shared risk and resilience framework requires a shared understanding of 5G security concerns and actions to increase security assurances, transparency, and accountability across Quad countries. Each measure must be evaluated on its merits and contribution to long-term security and resilience.

QUAD Quadrilateral Security Dialogue is an informal dialogue, a diplomatic and military arrangement modelled on the concept of Democratic Peace. Quad four member countries are—United States, India, Japan, and Australia.

The following is a risk and resilience framework for Quad countries that may also be used in the Indo-Pacific region as a whole. Three main risk categories are identified in the framework, which, if handled, will result in a complete strategy for safeguarding country 5G networks.

- **Technical Risk** refers to the dangers posed by 5G design and protocols.

- **Supply Chain and Connectedness Risk** addresses the threat of supply chain disruptions caused by natural and man-made factors such as conflict, geopolitical pressures, complex supply chain interdependencies, and state powers to regulate national suppliers.

- **Capability and Capacity Risk** addresses insufficient institutional skills and capacity, including in terms of expertise, institutional relationships, and staff to monitor and manage 5G security risks

However, new security risks and attack surfaces are introduced due to these additions and changes.

Technical Risk

Threats to each of the three "planes" of telecommunications architecture are mentioned as 5G security challenges like compromised user plane integrity, the additional complexity in the 5G control plane, and the security of the management plane. For instance, the massive number of endpoints—including IoT devices—could cause network traffic to increase unexpectedly.

Signaling storms, which are caused by app malware, can also bring down the entire network. Suppose user security parameters are not updated while migrating from one operator to another. In that case, 5G roaming may be challenging to secure. The line between "core" and "edge" networks has become hazier in 5G. Edge computing allows for low latency by lowering the time it takes for endpoints to communicate with the server by moving some key functions to the network's edge closer to the end-user. As a result, policy controls used in previous generations of communications to keep untrustworthy vendors and technologies out of the core are no longer effective in 5G. Vulnerabilities and vulnerabilities in 5G products, whether accidental or malicious, jeopardize 5G communications security. In the event of non-standalone (NSA) 5G designs, vulnerabilities from prior generations of wireless technologies can be passed down.

However, standalone (SA) 5G architecture may have yet-to-be-discovered flaws. Emerging solutions have their own set of hazards. With numerous countries deploying NSA 5G networks, there is a risk of establishing long-term vulnerabilities due to reliance on a single vendor. By providing interoperability and vendor diversity within a single network, the Open Radio Access Network (RAN) has been touted as a solution to avoid this lock-in. The flexibility of Open RAN, on the other hand, may introduce security risks, as the patchwork of components and interfaces makes them more vulnerable than traditional RAN architectures. As a result, each 5G technology option comes with its technological hazards. As part of a broader 5G security plan, various technical and risk assessments of network design, protocols, and components must drive policy and vendor selections based on international standards and industry best practices.

Supply Chain and Connectedness Risk

5G supply chains around the world are fundamentally risky. Hundreds of suppliers and subcontractors offer their components and services to 5G equipment makers, and components for 5G networking technologies are planned, developed, manufactured, assembled, and shipped worldwide.

The complexity of supply networks makes tracking and mitigating third-party risk from suppliers and contractors difficult and expensive, especially when they are located in foreign and/or unfriendly governments. As a result, it's challenging to keep track of third-party 5G equipment vendors and service providers.

In particular, tainted, counterfeit, and inherited software and hardware components can bring security vulnerabilities or hidden functions that threat actors can exploit in 5G technologies. This danger isn't confined to "untrustworthy suppliers" or "untrustworthy items." Even the security of well-engineered, reliable components from well-known vendors can be jeopardized. Suppose the designations "trusted solutions" and "trusted partners" aren't founded on open, international standards. In that case, there's a risk that the security baseline for 5G isn't raised high enough for all 5G manufacturers and operators. A "trusted supplier" breach might have far-reaching and potentially disastrous consequences. That concept was proved in the 2020 SolarWinds hack when a "trusted supplier" was the subject of a supply chain attack in which over 18,000 customers downloaded a malicious software update. Supply chain bottlenecks can cause supply disruptions or shortages. Natural disasters or man-made conflict at a specific link or region in the supply chain, a rise in global demand for 5G items, or even unrelated products that use the same component in large quantities or rely on the same production capacity might all contribute to this (e.g., semiconductor shortage across all sectors due to a surge in demand). Even a breakdown of the underlying logistics networks can cause costly interruptions, as demonstrated by a vast container ship blocking the Suez Canal in 2021. Finally, some states have attempted to weaponize the complex interdependencies of the 5G supply chain for strategic gains, such as in trade disputes, to apply pressure on trading partners. Concerns that present supply chain structures render markets vulnerable to the availability of crucial ICT (information and communications technology) components are driving a greater focus on innovation and industrial policy in liberal countries. COVID-19 also drew attention to a growing trend of protectionism. It's too early to know if similar dynamics will play out in the future, with countries restricting the sale of 5G components. Non-tariff barriers such as export licenses, internal taxes, new certification, product quality requirements, industrial policies, and the on-shoring of critical goods could all affect the availability of secure 5G components and delay the global deployment of 5G networks. Because there is no shared understanding of 5G supply chain risk and response mechanisms, states may impose arbitrary limitations that stifle innovation and halt the creation of a broad supply of internationally competitive, high-quality 5G components.

Risks of Capability and Capacity

While 5G offers better security features to the next generation of wireless communications networks like mutual authentication capabilities and enhanced subscriber identity protection, the deployment of 5G networks still poses several cybersecurity and policy problems. Capability and capacity across many governments and industry areas, including institutions, infrastructure, resources, and individuals, are required to manage and mitigate these risks successfully. Traditional telecommunications service providers (TSPs) aren't the only ones deploying 5G networks. Private entities, corporations, and institutions can run their own 5G networks. While large TSPs can secure their networks using industry standards, smaller participants may lack the operational maturity or resources to do so.

To combat 5G security concerns, institutional arrangements akin to national sector-specific Computer Emergency Response Teams (CERTs) are needed, and procedures and staff to monitor networks for intrusions like the Einstein & Continuous Diagnostics and Mitigation programs in the US federal government) are required. Collaboration between government bodies, TSPs, and industry is necessary to establish and sustain these risk competencies and capacities. For example, collaboration is required to develop and administer baseline security measures, incident response plans, and joint exercises. In addition, network operators must engage closely with industry partners and governments to respond to 5G security issues and exchange threat intelligence on time. Legal provisions to preserve essential infrastructure, allocated funds, or tax incentives can be used to build specified competencies and capacities. It's also worth mentioning that institutional capacity for dealing with 5G network vulnerabilities necessitates political capital and diplomatic skill.

Collaboration with trusted international partners must be supplemented by monitoring and assessment, which necessitates the creation of the necessary institutional "bridges," such as CERT-CERT cooperation and cross-pollination of cyber security standards for 5G deployments at the technological level, as well as statecraft and cyber diplomacy at the international level. Cohesion around precise definitions of risk aids in generating and maintaining political momentum for technical and institutional capacity building.

RECOMMENDATIONS

For managing risk and resilience in the 5G setup, a complete framework is required to provide an objective and transparent foundation for controlling and reducing threats and hazards, such as those posed by certain equipment and providers.

The measures for which the Quad countries may work together to create a shared risk and resilience framework have been simplified into five main proposals below. These recommendations should be used to address the three major risk categories. The weighting of the risk areas will be determined by the quad countries' experiences and priorities, which will likely change over time as the suggestions are implemented. Given their growing alignment on the subject of universal technology standards based on "similar interests and values," the Quad should:

1. Perform shared risk evaluations of 5G supply chains, including scenarios for common threat vectors and mitigation strategies for vendors and operators.

2. Establish similar standards for 5G vendors and equipment providers regarding "trustworthy" behavior.

At the moment, the four countries are working on their own initiatives, such as India's list of trusted vendors, which is overseen by the National Cyber Security Coordinator; Australia's review of existing telecommunications supply chain requirements, which is governed by the Parliamentary Joint Committee on Intelligence and Security; and the United States' threat assessment under the Enduring Security Framework, which is part of the Critical Infrastructure Partnership Advisory Council. Inconsistent standards make it difficult for vendors to comply with them, and they can stifle efforts to develop internationally competitive alternatives. As a result, nodal authorities in India, Australia, the United States, and Japan should convene a series of meetings to establish uniform criteria for trusted vendors, network operator requirements, and dangerous scenarios. These gatherings could put out several voluntary actions similar to those outlined in the European Union's 5G cybersecurity toolkit. For implementing proposals 1 and 2, Quad countries should collaborate closely with their respective national 5G industry players.

Leading chipmakers such as Qualcomm are based in the United States. Reliance Jio is aiding India's 5G technology development, and Japan has two local 5G equipment vendors—NEC and Fujitsu—that largely service the domestic market but have global ambitions.

The 5G Resilience Alliance might then extend the common standards that arise from this approach to other like-minded governments.

1. For the Quad Critical and Emerging Technologies Working Group, develop a 5G agenda. The recently formed Quad Critical and Emerging Technologies Working Group is one platform for developing a shared practice around 5G. Sharing 5G threat information, evolving joint security requirements for 5G procurement, adopting a standard testing and evaluation scheme for 5G security (e.g., Network Equipment Security Assurance Scheme), establishing reciprocity for certification, coordinating standard-setting activities, conducting joint 5G security exercises, and developing joint research agendas for 5G security and resilience are all measures that the Working Group could encourage.

2. Coordinate policy priorities on ICT security and standard-setting in international forums. Due to the large volume of data streaming through these networks, telecommunications networks will continue to be hotbeds for foreign intelligence activities, regardless of their capability and capacity to respond to 5G security issues. Technical security measures may increase the attacker's expenses, but sophisticated state actors will not be deterred. To strengthen international norms for responsible behavior in cyberspace and ensure the integrity of 5G infrastructure and supply chains, stronger international cyber norms, capacity-building efforts, and confidence-building measures developed through the United Nations and regional security and economic organizations are required. For example, the Group of Governmental Experts on Promoting Responsible State Behavior in Cyberspace report 2021 reaffirmed an earlier 2015 supply chain security standard, stating that "States should take reasonable steps to ensure the integrity of the supply chain so that end users can have confidence in the security of ICT products." States should work to prevent the spread of dangerous ICT tools and practices and the usage of potentially hazardous hidden features. Under the Global Commission on the Stability of Cyberspace's "Norm to Avoid Tampering," tampering with 5G network equipment that would systematically undermine security and significantly harm cyberspace

stability is likewise severely discouraged. Furthermore, the ITU and 3rd Generation Partnership Project (3GPP), GSMA(Global System for Mobile Communications),, O-RAN, and other important 5G standard-setting organizations and industry groups should pay particular attention to creating technical 5G standards. The attitudes and actions of the Quad's competent national authorities should be coordinated. The Quad countries should develop a complementary, if not joint, agenda for such international forums.

3. Create a multistakeholder 5G Resilience Alliance. The Quad cannot be the exclusive focus of efforts to create a resilient global 5G ecosystem. In this line, a multistakeholder 5G alliance to promote objective and transparent standards, cybersecurity, and other relevant specialists would benefit the ecosystem. The coalition should not duplicate initiatives elsewhere but build on existing relationships and agreements within and outside the Quad, such as the 2020 Japan-India ICT cooperation pact and the 2020 India-UK 5G MoU.

Taking a multistakeholder approach, the 5G resilience alliance should develop security and resilience measures relevant to and in close collaboration with organizations within the ICT industry and the larger ICT ecosystem, rather than limiting its efforts to state-centric actions. The following are some examples of feasible metrics at various levels: First, companies that buy and operate 5G equipment should examine third-party vendor risk and their risk management capabilities. This is to determine purchasers' risk-aware 5G procurement needs, which should adhere to recognized standards and best security practices. Buyers may demand that vendors follow secure development standards, deliver services and software in secure configuration by default, and implement security vulnerability management best practices. Second, industry requirements that apply to all 5G buyers and sellers can be set through market-driven collective purchasing power. To that goal, tax incentives or special funding could be used. Assurance, transparency, and accountability mechanisms should be employed across the supply chain to enhance the security baseline for all industry players. Transparency requirements for vendors, such as how they handle their ties with subcontractors and suppliers, would provide insights into supply chain dependencies and potential risks across industries. Third, policies that benefit both national and global 5G deployments are critical because the ICT ecosystem crosses national boundaries.

A 5G alliance could kick off efforts to create regional transparency and testing centers for code inspection and 5G compliance initiatives. Such centralized operations, such as Quad-wide testing methods and certification schemes, will enable cost-effective testing and one-time accreditation and verification to help speed up the 5G rollout timeframe.

KEY SECURITY CHALLENGES IN 5G

Because 5G will connect critical infrastructure, it will necessitate increased security to safeguard the safety of the critical infrastructure and the safety of society as a whole. For example, a security compromise in online power supply systems may be disastrous for all of society's electrical and technological systems. Similarly, we recognize the importance of data in decision-making, but what if crucial data is distorted while being sent across 5G networks? As a result, it's critical to look into and emphasize the significant security concerns in 5G networks and at the prospective solutions that could lead to safe 5G systems.

Network and Threat Landscape

The threat landscape for 5G consists of the following threats:

- The high number of end-user devices and new things cause a surge in network traffic (IoT).

- **Radio interface security:** Radio interface encryption keys are sent across insecure channels.

- **User plane integrity**: The user data plane has no cryptographic integrity protection.

- **Network security that is mandated:** Service-driven limits on the security architecture that lead to the voluntary adoption of security measures.

- **Roaming security:** When users roam from one operator network to another, their security parameters are not updated, resulting in security breaches.

- **Infrastructure denial of service (DoS) attacks**: visible network control elements and unencrypted control channels

- **Signaling storms:** Distributed control systems that require coordination, such as the 3GPP protocols' Non-Access Stratum (NAS) layer.

- **DoS attacks on end-user devices:** Operating systems, programs, and configuration data on user devices have no security safeguards.

SA WG3 of the 3GPP is actively involved in identifying security and privacy criteria and specifying security architectures and protocols for 5G. The Open Networking Foundation (ONF) is a non-profit organization dedicated to speeding the adoption of SDN (software Defined Networking) and NFV (Network Function Virtualization), by publishing technical specifications, including security specifications. Beyond radio efficiency, NGMN (Next Generation Mobile Networks)'s 5G design ideas include building a single composable core and simplifying operations and management through new computer and networking technologies. As a result, we concentrated on the security of the technologies that will comply with NGMN's design principles, such as mobile clouds, SDN, and NFV, and the communication links that will be utilized by or in between these technologies. Given the growing awareness about user privacy, we've also emphasized the possible privacy difficulties. The security problems are listed in Table 7.1.

Table 7.1 shows the many forms of security threats and attacks, as well as the targeted elements or services in a network, and other technologies that are most vulnerable to attacks or threats. The following sections provide a rapid overview of these security issues.

TABLE 7.1 Security Challenges in 5G Technology

Security Technology	Primary Focus	Target Technology				Privacy
		SDN	NFV	Channels	Cloud	
DoS, DDoS detection	Security of centralized control points	/	/			
Configuration verification	Flow rules verification in SDN switches	/				
Access control	Control access to SDN and core network elements	/	/		/	
Traffic isolation	Ensures isolations for VNFs and virtual slices		/			
Link security	Provide security to control channels	/		/		
Identify verification	User identity verification for roaming and clouds services					/
Identify security	Ensure identity Security of users					/
Location security	Ensure security of user locations					/
IMSI security	Secure the subscriber identity through encryption					/
Mobile terminal security	Anti-malware technologies to secure mobile terminals					/
Integrity verification	Security of data and storage system in clouds				/	
HX-DoS mitigation	Security for cloud web services				/	
Service access Control	Service-based access control security for clouds				/	

Mobile Cloud Security Challenges

Because cloud computing systems contain various shared resources, users can disseminate harmful traffic to slow down the entire system, consume additional resources, or gain unauthorized access to other users' resources. Interactions can also cause configuration disputes in multi-tenant cloud networks where tenants execute their control logic. Mobile Cloud Computing (MCC) integrates cloud computing concepts into 5G ecosystems. This leads to many security flaws, mainly caused by 5G's architectural and infrastructure changes. As a result, the MCC's open design and the adaptability of mobile terminals offer weaknesses via which adversaries might launch threats and compromise mobile cloud privacy. We can divide MCC risks into front-end, back-end, and network-based mobile security threats based on targeted cloud segments.

The client platform, which comprises the mobile terminal on which applications and interfaces are necessary to access cloud resources, is the MCC architecture's front-end. The threat landscape in this segment can range from physical threats, in which the mobile device and other integrated hardware components are the primary targets, to application-based threats, in which adversaries use malware, spyware, and other malicious software to disrupt user applications or collect sensitive user information. The cloud servers, data storage systems, virtual machines, hypervisors, and protocols necessary to provide cloud services make up the back-end platform. Security risks are mostly directed at the mobile cloud servers on this platform. These threats could include data replication to HTTP and XML DoS (HX-DoS) attacks. Radio Access Technologies (RATs) that connect mobile devices to the cloud focus on network-based mobile security concerns. This could be standard Wi-Fi, 4G Long Term Evolution (LTE), or other 5G-specific RATs. Wi-Fi sniffing, DoS/DDos attacks, address impersonation, and session hijacking are examples of this attack. Another significant area of interest in assessing the security challenges in 5G mobile clouds is the Cloud Radio Access Network (C-RAN). C-RAN can meet the industry's capacity expansion needs for increased mobility in 5G mobile communication networks.

On the other hand, C-RAN is vulnerable to the security risks associated with virtual systems and cloud computing technology. For example, C-centralized RAN's architecture risks becoming a single point of failure. Other dangers to the system include intrusion attacks, in which adversaries break into the virtual environment to monitor, modify, or run software routines on the platform while undetected.

SDN and NFV Security Challenges

SDN centralizes network control platforms and allows communication networks to be programmable. On the other hand, these two disruptive qualities open the door to network cracking and hacking. For example, DoS attacks prefer centralized control, and exposing important Application Programming Interfaces (APIs) to unwanted software can slow down the entire network. The SDN controller adjusts data path flow rules, allowing controller traffic to be easily identifiable. Because the controller is a visible object in the network, it is a popular target for DoS attacks. Because of saturation attacks, centralizing network control can make the controller a bottleneck for the entire network, as can be seen in. Because most network functions can be implemented as SDN apps, malicious applications that are given access to a network might cause havoc. Although NFV is critical for future communication networks, it faces fundamental security issues such as confidentiality, integrity, authenticity, and nonrepudiation. It is stated in that current NFV platforms do not provide adequate security and isolation to virtualized telecommunication services when used in mobile networks. The dynamic nature of VNFs(Virtual network functions), which leads to configuration errors and hence security failures, is one of the most persistent hurdles to adopting NFV in mobile networks.

Security Challenges in Communication Channel

Drones and air traffic control, cloud-based virtual reality, linked vehicles, smart factories, cloud-based robots, transportation, and e-healthcare will all be part of the 5G ecosystem. As a result, the applications require secure communication systems that can handle more frequent authentication and the transmission of more sensitive data. In addition, numerous new actors will participate in these services, including public service providers, mobile network operators (MNOs), and cloud operators. Several layers of encapsulated authentications are necessary at both the network access and service levels in such an ecosystem, and regular authentication between actors is required. Mobile networks used dedicated communication channels based on GTP and IPsec tunnels prior to 5G networks. Attacking communication interfaces such as X2, S1, S6, and S7, which are only utilized in mobile networks, necessitates a high level of skill. SDN-based 5G networks, on the other hand, will use standard SDN interfaces rather than dedicated interfaces. Because these APIs are open, the number of potential attackers will grow. The data channel, control channel, and inter-controller channel are the three communication channels in SDN-based 5G mobile networks. These channels are protected in the current SDN system by Transport Layer Security (TLS)/ Secure Sockets Layer (SSL) sessions. TLS/ SSL sessions, on the other hand, are highly vulnerable to IP layer assaults and SDN Scanner attacks and lack robust authentication techniques.

5G Privacy Challenges

Data, location, and identity could all pose serious privacy concerns from the user's perspective. Before installing most smartphone applications, the subscriber's personal information is required. The way data is saved and for what purposes it will be utilized is rarely mentioned by application developers or companies. Subscriber location privacy is mainly targeted by threats such as semantic information assaults, timing attacks, and boundary attacks. Access point selection methods in 5G mobile networks can expose location privacy at the physical layer. Catching assaults on the International Mobile Subscriber Identify (IMSI) of a subscriber's User Equipment can be used to reveal a subscriber's identity (UE). Such attacks can also be carried out by setting up a bogus base station that the UE(User equipment), recognizes as the preferred base station, causing subscribers to react with their IMSI. Remember that, Virtual MNOs (VMNOs), Communication Service Providers (CSPs), and network infrastructure providers are among the several actors in 5G networks. Security and privacy are different concerns for each of these actors. In a 5G network, synchronizing mismatched privacy standards across various parties will be a challenge. Mobile operators have direct access to and control over all system components in prior versions. However, as they will rely on new actors such as CSPs, 5G mobile carriers will lose complete control of the systems. As a result, 5G operators will lose full control over security and privacy. User and data privacy is seriously threatened in shared environments, where the same infrastructure is shared across many entities, such as VMNOs and other rivals. Furthermore, there are no physical barriers because 5G networks use cloud-based data storage and NFV characteristics. As a result, 5G operators have no direct control over where their data is stored in cloud settings. Because different countries have varied levels of data privacy protections based on their chosen context, the privacy of user data stored in a cloud in another country is jeopardized

Security Solutions with Potential

This section focuses on the solutions for the security issues discussed in the previous section. The problems of flash network traffic can be overcome by either adding more resources or enhancing the utility of current systems using cutting-edge technology. We believe that emerging technologies such as SDN and NFV can cost-effectively address these issues in a better manner. SDN enables the assignment of run-time resources, such as bandwidth, to specific areas of the network as the need arises. The controller in SDN can acquire network analytics from network equipment via the south-bound API to check if traffic levels are increasing. Services from the main network cloud can be moved to the edge using NFV to suit user requirements.

To deal with flash network traffic, virtual network slices can be dedicated only to locations with a high density of UEs. The security of radio interface keys remains a problem, necessitating secure key exchange, such as the proposed Host Identity Protocol (HIP)-based system. End-to-end encryption solutions can also be used to ensure user plane integrity. Centralized systems with worldwide awareness of users' actions and network traffic behavior, such as SDN, roaming security, and network-wide mandated security policies, can be achieved. Signaling storms will be increasingly complex because of the increased connectivity of UEs, small base stations, and high user mobility. C-RAN and edge computing are viable solutions to these problems, but their design must consider the increase in signaling traffic as a key feature of future networks, as stated by NGMN.

MANAGING DoS ATTACKS

The next sections describe solutions for DoS or saturation assaults on network control elements, one of the most common and disabling attacks on a network. The security solutions for the risks in the technologies discussed in the preceding part are listed below:

Mobile Cloud Security Solutions

The majority of MCC's recommended security solutions center on the strategic use of virtualization technologies, the redesign of encryption methods, and dynamic data processing point allocation. Because each end node connects to a specific virtual instance in the cloud via a Virtual Machine (VM), virtualization is a natural choice for protecting cloud services. This ensures security by isolating each user's virtual connection from that of other users. Similarly, service-based restrictions will ensure that cloud computing technologies are used safely.

For instance, the "Protect Sharing and Searching for Real-Time Video Data in Mobile Cloud" infrastructure uses cloud platforms and 5G technology to secure cloud services and allow mobile users to share real-time films on 5G equipped clouds. Unlike existing solutions that allow anybody with a shared connection to access Internet video feeds, our design only allows approved viewers access. Specific solutions, such as learning-based systems, are more valuable than general techniques for specific security threats like HX-DoS. To detect and mitigate risks, the learning-based system, for example, takes a given amount of packet samples and analyses them for certain known features. Anti-malware software could help safeguard mobile terminals and increase resistance to malware attacks. Anti-malware software is either installed on the mobile device or hosted and provided from the cloud. The security framework for MCC data and storage will include energy-efficient mechanisms for data and storage service integrity verification, as well as a public-proven data possession scheme and some lightweight compromise resilient storage outsourcing. For application security, some proposed frameworks are based on securing elastic applications on mobile devices for cloud computing, a lightweight dynamic credential generation mechanism to protect user identity, an on-device spatial cloaking mechanism for privacy protection, and MobiCloud, a secure cloud Framework for Mobile Computing and Communications.

C-RAN, a cloud-based architecture for RAN security, has been proposed for optimizing and providing safer RANs for 5G clouds. To meet this demand, C-RAN must offer a high level of reliability compared to traditional optical networks such as Synchronous Digital Hierarchy (SDH). One way to do so is through the widespread adoption of mechanisms such as fiber ring network protection, which are currently primarily used in the industrial and energy sectors.

SDN and NFV Security Solutions

SDN promotes speedy threat identification through a cycle of collecting intelligence from network resources, states, and flows, thanks to the logically centralized control plane with global network visibility and programmability. As a result, the SDN architecture facilitates network forensics, security policy changes, and security service insertion by supporting extremely reactive and proactive security monitoring, traffic analysis, and response systems. Due to global network visibility, security systems such as firewalls and

Intrusion Detection/Prevention Systems (IDS/IPS) can be used for specific traffic by updating the flow tables of SDN switches. The security of VNFs can be obtained by utilizing a security orchestrator in accordance with the ETSI (European Telecommunications Standards Institute) NFV architecture. The suggested architecture protects virtual functions in a multi-tenant environment and physical entities in a communications network. To provide hardware-based protection for private information and detect malicious software in virtualized settings, remote verification and integrity checking of virtual systems and hypervisors is proposed using trusted computing.

Communication Channel Security Solutions

To prevent the highlighted security concerns and retain the additional benefits of SDN, such as centralized policy management, programmability, and global network state visibility, 5G requires effective communication channel security. In today's telecommunication networks, such as 4G-LTE, IPsec is the most widely used security protocol for securing communication channels. With minor adjustments, IPsec tunneling may be used to protect 5G communication lines. Furthermore, several security techniques, such as authentication, integrity, and encryption, are integrated to ensure security for LTE communications. However, excessive resource usage, significant overhead, and a lack of coordination are the key issues in such existing security solutions. As a result, these technologies aren't suitable for 5G critical infrastructure connectivity. Thus, novel security mechanisms such as physical layer security using Radio-Frequency (RF) fingerprinting, asymmetric security schemes, and dynamically changing security parameters as per the scenario can be used to achieve a greater level of security for meaningful communication. End-to-end user communication can also be secured using cryptographic protocols like the Host Identity Protocol (HIP).

Note: The HIP is an internetworking architecture and set of protocols that was developed by the Internet Engineering Task Force (IETF). HIP improves the original Internet architecture by introducing a namespace between the IP layer and the transport protocols. The so-called identifier/locator split is implemented in this new namespace, which is made up of cryptographic identifiers.

Safety and Security Solutions for Privacy in 5G

5G must have privacy-by-design approaches. Privacy is considered from the start of the system as many essential elements are built-in. A hybrid cloud-based strategy is needed for safeguarding the privacy. Mobile operators can store and process highly sensitive data locally while storing and processing less sensitive data in public clouds. Operators will have more access to and control over data and will be able to determine where to share it. Similarly, with 5G, service-oriented privacy will lead to more realistic privacy solutions. Better methods for accountability, data minimization, transparency, openness, and access control will be required for 5G. As a result, robust privacy protections and legislation should be considered throughout the standardization of 5G. There are three different sorts of regulatory approaches. The first is government regulation, which mostly consists of country-specific privacy legislation enacted by governments and organizations like United Nations (UN) and the European Union (EU). The second level is the industry level, where diverse industries and organizations such as 3GPP, ETSI, and ONF collaborate to draft the best privacy principles and practices. Third, rules at the consumer level ensure that desired privacy is protected by taking into account customer needs. For location privacy, anonymity-based approaches must be used, in which the subscriber's real identity is masked, and pseudonyms are used instead. In this circumstance, encryption-based procedures are also helpful; for example, messages can be encrypted before being sent to a Location-Based Services (LBS) provider. Obfuscation techniques are also beneficial in which location information quality is decreased to safeguard location privacy. Furthermore, spatial cloaking-based algorithms are beneficial for dealing with some of the most common location privacy threats, such as timing and boundary attacks.

Note: Making anything challenging to grasp is referred to as obscuring. Programming code is frequently obfuscated to safeguard trade secrets or intellectual property and stop an adversary from deciphering a proprietary software application.

Since LBS is provided to users based on their precise location information, a major threat to user privacy has been reported. Spatial cloaking has recently been frequently employed to combat such privacy infringement in LBS.

The fundamental principle behind the spatial cloaking approach is to conceal a user's precise position within a cloaked area while still maintaining the privacy standards that the user has selected.

Note: Spatial cloaking is a technology that blurs a user's precise position into a spatial zone to protect her/his privacy. The user's stated the blurred spatial region must meet privacy criteria.

K-anonymity and minimum spatial area are the most commonly employed privacy requirements. The condition of k-anonymity ensures that a user's location is unidentifiable among k other users. The minimum spatial area criterion, on the other hand, ensures that a user's particular position is blurred into a spatial region with an area of at least A.

CHAPTER 8

Security for 5G Mobile Wireless Networks

Passive and active security attacks are the two sorts of attacks that can be classified. Attackers who utilize a passive attack aim to learn or use information from legitimate users but, do not intend to attack the communication itself. Eavesdropping and traffic analysis are the two most common passive attacks in a cellular network. Passive attacks are designed to compromise data security and user privacy. In contrast to passive attacks, active attacks might involve data manipulation or the disruption of authorized communications. Man-in-the-middle (MITM) attacks, replay attacks, denial of service (DoS) attacks, and distributed denial of service (DDoS) assaults are all examples of active attacks.

Cryptographic techniques using new networking protocols and physical layer security (PLS) approaches are the primary mechanisms to combat security attacks. Cryptographic approaches are the most often utilized security mechanisms, typically applied at the upper layers of 5G wireless networks using new networking protocols. Strict-key cryptography and public-key cryptography are the two types of modern encryption. Symmetric-key cryptography refers to encryption technologies in which a sender and receiver share a secret key.

Asymmetric encryption, in contrast to "regular" (symmetric) encryption, encrypts and decrypts data using two distinct but mathematically related cryptographic keys. Those are a Public Key and a Private Key.

DOI: 10.1201/9781003264408-8

They are together referred to as a "Public and Private Key Pair." The critical length and computational complexity of the algorithms determine the performance of a security service. In older cellular networks, the administration and distribution of symmetric keys are effectively safeguarded.

A cryptographic algorithm sits at the core of asymmetric encryption. This approach creates a key pair using a key generation protocol, which is a type of mathematical function. The way the keys are related to one another varies depending on the algorithm.

The management and distribution of symmetric keys may face significant issues in 5G due to increasingly complicated protocols and varied network architectures. The application of PLS has been restricted due to limited progress on practical wiretap codes and strictly positive secrecy capabilities in the 1970s and 1980s. At the time, public-key cryptography was used in most modern security techniques. After demonstrating that a legitimate user with a worse channel than the eavesdropper may nonetheless produce a secret key over an insecure public channel, interest in employing PLS immediately grew. In 5G wireless networks, there have been many PLS studies recently. PLS has been recognized as a viable security strategy for providing safe wireless communications using the unique wireless physical layer medium properties, unlike traditional systems that primarily provide security through cryptographic techniques. PLS has two distinct benefits over cryptography: low computing complexity and high scalability, making it a viable candidate technique for cryptographic key distribution in 5G wireless networks. According to its theoretical security capacity, power, code, channel, and signal approaches, Shiu et al. summarized the available PLS techniques and categorized them into three essential categories. Aside from PLS and cryptographic methods, some research has been done on security architecture, vulnerability assessment tools, and data-driven intrusion detection mechanisms.

These security measures must meet 5G performance criteria, including ultra-low latency and a high degree of energy efficiency (EE). As a result, legacy security features, new use cases, and new networking paradigms must be included in the 5G security requirements. The typical features of a 5G security architecture are shown in Figure 8.1. Edge cloud is used to boost network performance by lowering communication latency. The central cloud connects the edge clouds for data sharing and centralized control.

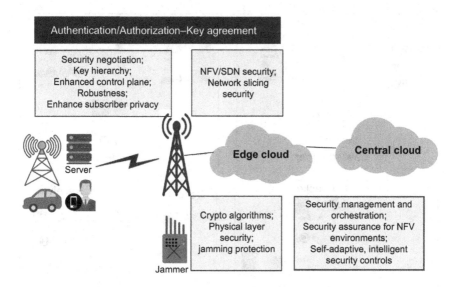

FIGURE 8.1   Elements in a 5G security architecture.

ATTACKS AND SECURITY SERVICES IN 5G NETWORKS

Wireless information transmission is exposed to various malicious risks due to the broadcast nature of the wireless medium. This part looks at four forms of attacks in 5G wireless networks: eavesdropping and traffic analysis, Jamming, DoS and DDoS, and MITM. Among the four security services we introduce are authentication, confidentiality, availability, and integrity.

Attacks in 5G Wireless Networks

Each attack is discussed separately in the following three aspects: type of attack (passive or active), security services provided to combat this attack, and the corresponding methods used to avoid or prevent this attack (see Figure 8.2). We will concentrate on security threats at the PHY (Physical) and MAC (Medium Access Control) layers, where significant security disparities between wireless and wired networks occur.

Eavesdropping and Traffic Analysis

Eavesdropping is an attack in which an unwanted receiver intercepts a message. Eavesdropping is a passive attack because it does not affect everyday communication, as demonstrated in Figure 8.2a. Eavesdropping is challenging to detect due to its passive nature. The most popular method

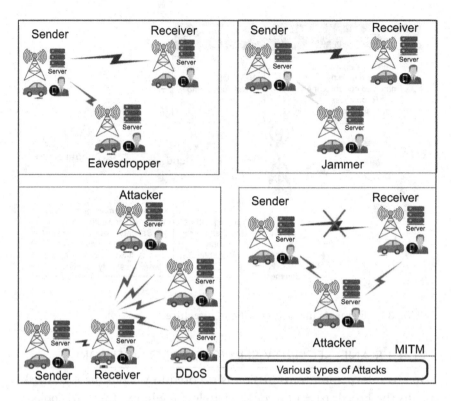

FIGURE 8.2 Attacks in 5G wireless networks. (a) Eavesdropping, (b) Jamming, (c) DDoS, and (d) MITM.

of preventing eavesdropping is to encrypt the signals sent over the radio channel. Because of the implementation of encryption, an eavesdropper cannot directly intercept the receiving signal. Another passive attack is traffic analysis. An unintentional receiver analyses the traffic of a received signal without understanding the signal's content to capture information such as the position and identities of the communication parties. In other words, even if the signal is encrypted, traffic analysis can still be utilized to reveal the communication parties' patterns. The exemplary communications are also unaffected by the traffic analysis assault.

The encryption method used to prevent eavesdropping relies on the encryption algorithm's strength and the eavesdropper's computer capability. Eavesdroppers can use new technology in their attacks due to the rapid escalation of computing power and the blossoming of advanced data analysis technologies. Existing anti-eavesdropping techniques are complicated because many presume a small number of simultaneous eavesdroppers with limited computation and data analysis capabilities. Furthermore, some 5G wireless network technologies, such as HetNet, may make it much more challenging to combat unauthorized listeners.

In general, the new characteristics of 5G wireless networks create a plethora of more challenging scenarios for dealing with eavesdroppers, such as when many antennas are considered. Because cryptographic methods to combat eavesdropping have been thoroughly explored in the past and are now deemed mature, PLS research to combat eavesdropping has recently received increased interest.

Jamming

Unlike eavesdropping and traffic analysis, Jamming can entirely interrupt legitimate user communications.

An example of a Jamming attack is shown in Figure 8.2b. The malicious node can cause intentional interference, causing legitimate users' data communications to be disrupted. Jamming can also make it impossible for an authorized user to access radio resources. Active attack solutions are typically detection-based. Spread spectrum techniques like direct sequence spread spectrum (DSSS) and frequency-hopping spread spectrum (FHSS) are frequently employed to combat Jamming at the PHY layer by spreading the signals over a broader spectral bandwidth. Anti-Jamming techniques based on DSSS and FHSS, on the other hand, may not be suitable for specific applications in 5G wireless networks. A pseudorandom time-hopping anti-Jamming strategy for cognitive users is suggested to increase performance over FHSS. Detection is possible due to the features of Jamming. A resource distribution strategy between a fusion center and a jammer is proposed. Resource allocation is used to improve detection and improve the performance of error rates.

DoS and DDoS

DoS attacks can deplete an adversary's network resources. DoS is a network availability violation caused by a security attack. A DoS attack can be launched via Jamming. When there are multiple distributed adversaries, a DDoS can be produced. A DDoS model is shown in Figure 8.2c. Both DoS and DDoS are active attacks that can be used at several layers. Detection is now used to identify DoS and DDoS attacks. DoS and DDoS will undoubtedly become highly dangerous for operators as the adoption of multiple devices in 5G wireless networks increases. In 5G wireless networks, DoS and DDoS attacks can use many connected devices to attack the access network. Depending on the attacking target, a DoS attack can be classified as either a network infrastructure or a device/user DoS attack.

DoS attacks on network infrastructure can affect the signaling plane, user plane, management plane, support systems, radio resources, logical and physical resources, etc. DoS attacks on devices and users can target battery, memory, disc, CPU, radio, actuators, and sensors.

MITM

In a MITM attack, the attacker secretly takes control of a communication channel between two legitimate parties. The MITM attacker can intercept, change, and replace communication messages sent between two valid parties. A MITM attack is depicted in Figure 8.2d. MITM is a type of active assault that can be launched at several levels. Data security, integrity, and availability are specifically threatened by MITM attacks. According to Verizon's data investigation report, the MITM attack is one of the most common security threats. False base station-based MITM is an attack in which an attacker compels a genuine user to establish a connection with a fake base transceiver station in a legacy cellular network.

Note: Mutual authentication between the mobile device and the base station is usually utilized to prevent false base station-base.

SECURITY SERVICES IN 5G WIRELESS NETWORKS

The new architecture, technology, and use cases of 5G wireless networks introduce new security features and needs.

Authentication

Entity authentication and message authentication are the two types of authentication. 5G wireless networks require both entity and message authentication to combat the attacks mentioned above. Entity authentication verifies that the communicating entity is whom it claims to be. Once the two parties can communicate, mutual authentication between user equipment (UE) and mobility management entity (MME) is required in older cellular networks. In the classic cellular security framework, mutual authentication between UE and MME is the most significant security element. Asymmetric-key authentication and key agreement (AKA) is used in 4G LTE cellular networks. 5G, on the other hand, necessitates authentication between the UE and the MME and third parties such as service providers. In 5G, hybrid and flexible authentication management is required since the trust paradigm changes from that employed in existing cellular networks. UE hybrid and flexible authentication can be implemented in three days: network-only authentication, service-only authentication, and network-and-service-provider authentication. Authentication in 5G wireless networks is predicted to be significantly faster than ever because of the extremely high data rates and low latency requirements of 5G wireless networks.

Furthermore, in 5G, the multi-tier architecture may experience frequent handovers and authentications between different levels. An (Software defined networking) SDN-enable fast authentication scheme based on weighted secure-context-information transfer is proposed to address key management challenges in HetNets and reduce the unnecessary latency caused by regular handovers and authentications between different tiers. It improves authentication efficiency during handovers and meets the 5G latency requirement. A public key-based AKA is presented to provide better security services in 5G wireless networks.

Message authentication is becoming increasingly critical in 5G wireless networks due to the different uses; moreover, 5G's stricter latency, spectrum efficiency (SE), and EE standards. A Cyclic Redundancy Check (CRC)-based message authentication system for 5G is proposed to detect random and intentional errors while reducing bandwidth usage.

Confidentiality

Data confidentiality and privacy are the two characteristics of confidentiality. Data confidentiality safeguards data transmission against passive attacks by restricting data access to only authorized users and prevents access or disclosure to unauthorized users. Privacy protects legitimate users' information from being controlled and influenced by an attacker. For instance, privacy shields against an attacker's analysis of traffic flows. The traffic patterns can be utilized to diagnose sensitive data, such as the location of senders and receivers, and so on. Massive data linked to user privacy exists in numerous 5G applications, such as car routing data, health monitoring data, etc. Data encryption is usually used to protect data confidentiality by prohibiting unauthorized users from extracting relevant information from broadcast data. The symmetric critical encryption approach encrypts and decrypts data using a single private key shared by the sender and receiver. A secure key distribution mechanism is necessary to communicate a key between the sender and the receiver. The premise behind traditional cryptography methods is that attackers have limited computing power. As a result, combating attackers with high computer capabilities is difficult. PLS can offer confidentiality services against Jamming and eavesdropping attacks rather than relying simply on general higher-layer cryptographic techniques. Users are beginning to recognize the necessity of privacy protection services and 5G data offerings. Because of the vast data connections, 5G privacy demands significantly more attention than conventional cellular networks. In many user scenarios, anonymity is an essential security requirement. Privacy breaches can have significant repercussions in several circumstances. Health monitoring data, for example, can reveal sensitive personal health information, while vehicle routing data can compromise location privacy. Privacy leakage is a significant risk with 5G wireless networks. The association method in HetNets can betray users' location privacy due to the high density of small cells. To ensure location privacy, a differentially private algorithm is presented. The suggested protocol protects the privacy of group communications.

Cryptographic algorithms and schemes are presented to provide secure and privacy-aware real-time video reporting service in-vehicle networks.

Availability

The availability is the degree to which a service is accessible and usable by lawful users whenever and wherever it is requested. Availability assesses the system's resilience in various threats and is a critical performance parameter of 5G. A typical active attack is like an availability attack. DoS/SSoS (Standards Setting Organizations), attacks are one of the most commonly used attacks on availability, and they can prevent genuine users from accessing services.

Jamming or interference may interrupt legitimate users' communication links by interfering with radio signals.

5G wireless networks have significant difficulty preventing Jamming and DDoS assaults due to many unsecured IoT devices. DSSS and FHSS are two traditional PLS methods for PHY availability. In the 1940s, the DSSS was initially used in the military. In DSSS, the spectrum of the original data signal is multiplied by a pseudo-noise spreading code. A jammer needs more power to disrupt genuine transmissions if they don't know about pseudo-noise propagation coding. A signal is conveyed via FHSS by rapidly switching between numerous frequency channels using a pseudorandom sequence created by a shared key between the transmitter and receiver. To improve the SE in 5G, dynamic spectrum is applied to D2D (Direct Device-to-Device) communications using the cognitive radio paradigm. FHSS, according to Adem et al., can degrade performance during a Jamming attack. A pseudorandom time-hopping spread spectrum is proposed to increase Jamming, switching, and mistake probability. To improve the identification of availability violations, resource allocation is used.

Integrity

Message authentication ensures that the message's source is verified, offering no security against duplicates or changes. 5G aspires to deliver connectivity anytime, anywhere, and anyhow and assist applications closely tied to everyday human life, such as water quality metering and transit schedules. In some applications, data integrity is one of the most important security requirements.

The term "integrity" refers to the ability to prevent data from being modified or altered due to active attacks by unauthorized parties. Insider malicious attacks can compromise data integrity, such as message injection or data alteration attacks. Insider attacks are challenging to detect because insider attackers have legitimate identities. Data integrity services must be given against tampering in use cases such as smart meters in the smart grid.

Data can be more easily attacked and manipulated than speech conversations. Mutual authentication, which generates an integrity key, can be used to deliver integrity services. It is necessary to provide a service that ensures the integrity of personal health information. The authentication systems can provide message integrity.

THE SECURITY FOR TECHNOLOGIES APPLIED TO 5G WIRELESS NETWORK SYSTEMS

We'll discuss health testing techniques from a 5G security standpoint in this section. First, we'll go over the 5G platform fast. Then the security practices of each technology are examined. HetNet, MIMO, D2D, SDN, and IoT infrastructure are employed in this section for 5 G broadband networks.

Het Net

CoMP (coordinated multipoint processing) can be utilized in HetNet to improve contact coverage. As a result of CoMP, the likelihood of legitimate users being eavesdropped on will increase. Several BSs (base stations), are picked to deliver the message. The proposed hierarchical BS classification approach is based on the possibility of scope protection. Based on theoretical and simulation testing, the authors determined that the proper BS selection level for CoMP will boost the secure coverage efficiency. Security-based resource management was utilized to apply protection in HetNet. Wang et al., with the help of SBSs (small-cell base-stations), made a case for strengthening present Jamming and transmitting systems by proposing a cross-layer collaboration to protect the secrecy of cellular macro communications. Through a monetary or resource incentive, SBSs are motivated to jam secure communications while adhering to the QoS of their users. The methodology based on intrusion detection is regarded as a secure method of communication. Intrusion prevention systems for virtual cloud computing are developed using heterogeneous 5G. Signature-based detection, irregular detection, specification-based detection, static configuration review, and hybrid intrusion detections based on these methods are among the methodologies examined. Various security standards address standard password protection and biometric verification.

D2D

D2D node cooperation is a systematic method of preventing eavesdroppers from listening in on D2D communication. Legal transmitters with a shared receiver will cooperate to maximize their effective broadcast efficiency. To enable distance-considered D2D communications, Ghanem and Ara proposed a collaboration agreement. Before collaborating, devices should test the gap to see if cooperation will improve communications security until the contract. To maximize the confidentiality levels achieved, range boundaries should be utilized to mutually define cooperation, cooperation from one side, or no coordination. Without any specific criterion for D2D communications, the new framework should be used for any D2D communications circumstances. Considering the network's support, the main communication protocols for the two D2D users and eNodeB are introduced. There are two options available. In the traffic offload scenario, D2D users are linked to the same eNodeB. D2D connections are required for apps in each D2D consumer for the social working method. Key sharing is accomplished through the use of public and authenticated dedicated networks. eNodeB is used for initial vital exchanges and mutual authentication of D2D users. Based on the purpose of eNodeB, three separate key exchange protocols with different computational times and sophistication are suggested throughout the authentication method.

Massive MIMO

On the other hand, eavesdroppers can utilize massive MIMO to attack lawful transmissions. Both BS and the eavesdropper found large MIMO with the Chen et al. device architecture. The antenna clusters of the eavesdropper are much more significant. This is where the OSPR strategy is introduced. According to theoretical and simulation analyses, the BS antenna number may have a considerable safety effect. Because of the sufficient number of antennas at the BS, the huge MIMO Eavesdropper will be unable to read most original symbols; however, legal users will be able to retrieve original characters with only a limited number of antennas. Compared to existing Jamming strategies, the proposed process has a greater EE.

SDN

SDN facilitates unified network access by separating the ground plane from the data plane and provides unique network administration methods such as network simplification, configuration, and flexibility. The SDN can be used to transfer information between cells. The SDN should have three fundamental qualities, namely logically unified knowledge, programmability, and abstraction, to improve significantly network scalability and efficiency while lowering costs. A survey based on mobile network software (SDMN) and related security concerns is offered. Dabbing et al. discuss the advantages and disadvantages of SDN defense. The benefits of SDN security over traditional networks are shown in Table 8.1. The table

TABLE 8.1   The Pros of SDN Security Over Traditional Network

SDN Characteristic	Attributed To	Security Use
Global network view	Centralization traffic statistics collection	Network-wide intrusion detection, Detection of switch's malicious behavior network forensics
Self-healing mechanisms	Conditional rules traffic statistics collection	Reactive packet dropping Reactive packet redirection
Increased control Capabilities	Flow-based forwarding scheme	Access control

shows the most recent security issues caused by SDN compared to the SDN benefits introduced to cellular 5G networks and any necessary solutions.

IoT

A two-hop randomly dispersed transmission should be used to protect relay transmission in IoT networks, considering power supply and code-word rate design. The question is posed to maximize the secrecy limit. Relays and eavesdroppers have single and multiple antenna instances. By increasing the number of antennas at relay knots, it is shown that proper relay communication will expand secure coverage and boost protection rates. RFID (Radio Frequency Identification) is a widely utilized IoT network technology for autonomous data collection and identification. A secure RFID reversal system is presented to employ multi-application RFID and revoke tag-related applications efficiently. The proposed program will achieve a higher level of safety than present systems based on the theoretical study.

CHAPTER 9

Security Risk Prevention and Control Deployment for 5G Private Industrial Networks

The installation of network infrastructures has expedited the commercialization of 5G networks. Different sectors have distinct characteristics and security requirements due to differences in business models and application scenarios. The focus of private network building has shifted to ensuring network security for industrial customers with secure 5G networks. Security concerns have been cited as a significant impediment to the growth of the Internet of Things (IoT).

Enhanced Mobile Broadband (eMBB), Massive Machine Type Communication (mMTC), and Ultra-reliable and Low Latency Communications (URLLC) are three application scenarios that 5G networks are expected to enable.

The fundamental components of a 5G network are the access network (e.g., base stations, access points, edge computing servers), much as they were in earlier generations of mobile networks. Private industrial networks can now be supported by 5G networks, a new functionality. The 5G network infrastructures are specifically deployed for diverse private industrial networks in a way that collaborates with various industry application

DOI: 10.1201/9781003264408-9

127

platforms to serve a variety of application scenarios. 5G application scenarios, on the other hand, are characterized by high complexity, variability, and unpredictability due to complex human-to-human, human-to-things, and things-to-things interactions.

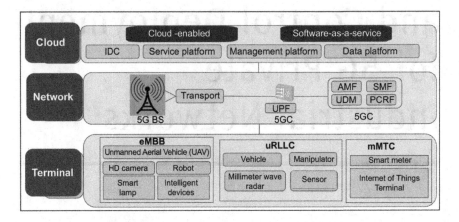

FIGURE 9.1   The architecture of 5G industry network.

5G introduces a novel network slicing function, i.e., an end-to-end virtual network that allows wireless access, transmission, and core networks to be virtualized from end to end. The network slicing feature securely divides 5G network resources while simultaneously providing services to consumers from various industries and businesses. Network Function Virtualization (NFV) and Software-Defined Networking (SDN) are used to build such a 5G network. Cloud computing services, in particular, may be realized on a universal hardware platform thanks to the virtualization of both software and hardware.

The 5G network standards addressed the security flaws in the 4G network and ensured that communications were secure.

The following elements are included in the security design of 5G networks:

i. Network Access Security: a set of security measures requiring the terminal to securely authenticate and access network services to safeguard against attacks on radio interfaces, particularly the wireless interface. 5G AKA is used for authentication in 5G standards. 5G networks, in particular, must handle a wide range of real-time applications, such as video streaming and vehicle-to-everything (V2X) communications, while also meeting the applications' stringent quality of service (QoS) requirements.

ii. Network Domain Security: a set of security features that allows network nodes to securely exchange signaling and user plane data and enable the network interface to transmit 256-bit keys.

iii. Service-based Architecture (SBA): a set of security features that assures the 5G SBA architecture's network functions can safely interface with other network domains, including network function registration, discovery, and authorization security.

iv. Security visibility: a set of security features that provides a set of functions that allow users to see if the safety function is active.

5G NETWORK CONSTRUCTION FOR SECURITY RISK PREVENTION AND CONTROL

The building of 5G networks is based on SBA, and network hardware is moving toward cloud-based NFV. Meanwhile, 5G is predicted to provide a $1000 \times$ data throughput, driving up demand for industrial systems that are open, interconnected, and intelligent. A recent poll found that large and medium-sized industries, such as transportation, electric power, medical care, and industrial parks demand openness, interconnectedness, and intelligence more. Private industrial networks should follow the following guidelines to meet the security requirement:

i. All application data needs to be preserved within the corporate park: existing methods can be used to maintain all application data, and data transmission within the park is prohibited.

ii. Network security must be ensured by the 5G basic network. Traditional public networks have gradually led to private networks in corporate security operations. The risk of data leakage is reduced if application data is maintained within the industrial park, and network attacks may be fought and managed.

SECURITY RISKS FACED BY 5G PRIVATE INDUSTRIAL NETWORK

Many company scenarios can benefit from a 5G private industrial network. Figure 9.1 depicts a typical 5G industry network design.

i. Terminals: In various service scenarios, numerous types of access terminals are used. One sort of terminal is sensing and monitoring equipment, such as millimeter-wave radar, multiple sensors, HD cameras, and so on, which collect data from the environment and communicate it to the network in raw form. The other form of the terminal is an application terminal, which may accept and execute commands and tasks provided by applications, such as a vehicle, an automated guided vehicle (AGV), a smart meter, etc.

ii. Network: 5G proprietary network consists of 5G base stations, access networks, and core networks.

iii. Cloud: A 5G private network's application layer that supplies and implements hardware and software for computing services.

The critical security needs of a 5G private network are as follows, based on the unique features of the three 5G application scenarios (eMBB, mMTC, and URLLC):

i. Security considerations in an eMBB scenario

High-data-rate applications, such as HD video and VR/AR [AR (augmented reality), VR (virtual reality)], are prevalent in this situation. The service's security requirement is that the 5G network reliably sustains transmissions with high data rates and high traffic loads, i.e., without or with minimal interruption. For quick network recovery, this needs tight criteria.

ii. Security considerations for the mMTC scenario

A significant number of IoT devices characterize this scenario, and typical industrial applications include smart grids and smart water conservation. The following are the security requirements:

First, application data security: Massive data must be transferred and processed safely.

Second, given the many terminals spread across a given area, it is necessary to ensure that their access processes are secure.

Security vulnerabilities caused by terminals and security threats to the network and cloud can be avoided, focusing on preventing DDoS (Distributed denial of service) attacks.

Third, network disaster recovery security is anticipated to safeguard or relocate application data and execute some level of self-recovery in unforeseeable natural catastrophes that cause significant damage to network infrastructures.

iii. URLLC scenario security needs

In this scenario, terminals are required to process applications with high latency requirements, such as autonomous driving, remote medical care, etc. The following are the security requirements:

First and foremost, application security. It is vital to avoid security concerns and ensure that varied applications are reliable.

Controlling safety is the second step. While using a range of authentication methods, it is vital to ensure the control and security of many types of terminals in various application scenarios.

The third point to consider is network security. To ensure the safety of the 5G private network without interruption, it is required to provide a highly reliable connection to diverse applications.

To summarize, network security is the primary concern in 5G scenarios. Customers from various industries are interested in learning how the 5G private network can fully guarantee network security while avoiding the impact of network security flaws on the enterprise cloud.

In private enterprise clouds, network security threats are generally related to terminal access, open edge networks, and internal border control.

i. Terminal tempering

USIM (Universal Subscriber Identity Module) cards are included with 5G access terminals and can be placed into an unauthorized terminal. Suppose a terminal like this is linked to the network. In that case, it can launch network assaults, putting the 5G network and the business intranet in danger.

ii. Open edges networks

Edge computing is a critical technique for industry customers/terminals to connect to 5G networks, but its open interface poses security and legal problems. As a significant part of the core network, Edge computing has altered network architecture and business models. Edge computing exposes the network barrier to third-party apps and the Internet, posing security dangers to the underlying cloud infrastructure and third-party applications.

iii. Securing the enterprise private cloud's perimeter

The private virtual layer in a corporate, private cloud may be subject to unauthorized tampering or infected with viruses or Trojan horse malware. Such dangers could jeopardize the security of 5G networks and lead to data leakage.

POTENTIAL 5G PRIVATE INDUSTRIAL NETWORK SECURITY RISKS

5G private industrial networks must regularly adapt to 5G security technologies. The IT cloud security and control requirements can be met, and the security and control of the "cloud network end." Two aspects of safety and control have been improved and are as follows.

i. Improvement in network security supervision: implement hierarchical access management and control for all terminals participating in private network communication. Edge computing can improve internal and external security monitoring with edge security opening capabilities, while various access terminals can improve access management. Security threats can be successfully reduced by strengthening the security and control of access terminals.

ii. Improvement in the isolation method for physical security: Network slicing, which combines security management with internal and external network security and control, is used to establish soft channel isolation. This is part of strengthening the hard isolation between cloud resource pools and cloud network defenses at the physical layer. This approach entails improving the isolation of cloud firewall hardware and access control to end various unexpected threats.

NETWORK SCHEME FOR 5G PRIVATE INDUSTRIAL NETWORK SECURITY

Security and control are becoming increasingly crucial as 5G commercial applications proliferate. For example, the level of protection in a 5G intelligent power system is higher than in the traditional power business. As a result, risk prevention and control in actual network implementation must be investigated.

In building a 5G private network, existing concepts were used to deploy firewalls, with internal and external isolation imposed between the data carrier network and the power industry network.

It is vital to concentrate on increasing the security and control of access terminals to improve network security supervision. 5G terminal security access technology is primarily utilized to secure identity authentication and access control in a system with many terminals. Edge computing access hazards induced by card manipulation and the resulting 5G network security vulnerabilities are removed in the existing machine-card separation networking configuration.

As a result, a two-level terminal authentication system, such as the 3A system, should be installed in private industrial networks.

Access Control via Two-level Terminal Authentication

5G power CPE (Customer Premise Equipment) will be utilized in 5G networks to address the need for anti-interference and equipment safety. Power terminals will be deployed in many 5G smart grid application scenarios.

Implementing 5G access control and deploying a two-level authentication (3A) system based on network authentication for terminal user cards is crucial for risk prevention. The 3A authentication system ties the user number to the access terminal's username, password, and IMEI and then performs access authentication. Customers in a specific industry can rely on the authentication process to meet their management needs while also ensuring access security. Figure 9.2 depicts the design of a private network for a 5G electric power system.

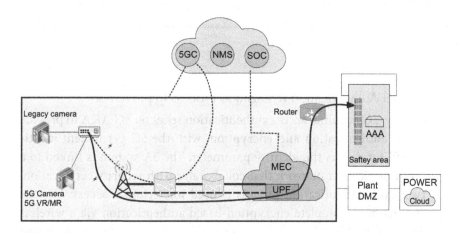

FIGURE 9.2   Diagram of 3A certification system.

In 5G edge computing systems, the A3A authentication method can be used. Alternatively, the authentication server can be installed in an industry network's DMZ. Table 9.1 shows a comparison.

The authentication procedure in a natural system is implemented at the network edge, which is supported by edge computing servers' computational and storage resources to improve the security of edge computing and 5G networks.

As a result, the service delay will be kept to a minimum, and the delay requirement is no more than 20 ms due to the two-level authentication exchange.

TABLE 9.1   Comparison of Different Deployment Methods

Deployment Comparison	Edge Computing Servers	Industry Intranet
Convenience	Directly provided by the deployment of security domains when edge computing is integrated	It needs to add a 3A authentication server
Process Security	Developed in the 5G network edge to reduce information forwarding Simultaneously enhance the security of edge computing and subsequently establish an edge computing security system	The intranet belongs to the enterprise network isolation area with unknown risks
Operation and Maintenance	The network operator coordinates operation and maintenance	The ownership of assets, operation, and maintenance needs to be negotiated between both parties: enterprise and network operator

Two-level Authentication Realization Process Improvement

Currently, 5G terminal access authentication relies on 5G AKA to provide two-way authentication and encryption, with the 5G permanent identifier SUPI serving as the identity parameter. The 3A system is linked to a 5G communication network that collaborates with security certification. A two-level authentication process is used for terminal access: Various endpoints must complete encryption-based authentication via a wireless network and a User Data Management (UDM) system before accessing the 5G network. Once a new 3A authentication mechanism is started, two-level authentication is done between the terminal and the 3A authentication server.

First, SMF (Session management function) takes user identity information, i.e., the SUPI, from the UDM system and examines whether the SUPI value matches the user type before deciding whether or not access should be permitted. The second step starts only if the first stage of the authentication procedure is successful. The following is the second phase in the authentication process:

Authentication trigger: PAP/CHAP authentication is used to get authentication information from DN-AAA by the SMF that has authorized the user.

A combination of user number, PEI (similar to IMEI), user name, and password can be used to verify the information. UPF (user plane function) creates a channel for authentication: The two-level authentication server must be deployed outside the 5G private network. The UPF must establish a PDU data channel to convey authentication information according to the requirements. SMF maintains the following authentication data: After the authentication is complete, the SMF can save the DNN name of the 3A authentication server to make obtaining the second authentication easier and limit the number of times the process is repeated. Because industry terminals do not now support the EAP protocol established by 3GPP, the 3A Radius protocol is used in the initial build. The core network SMF to 3A multiplexes the data bearer network's signaling plane channel in smart power generation and power grid monitoring scenarios. IPsec encrypted tunnels are utilized for information transmission since the communication includes sensitive information.

TABLE 9.2   Parameter and Settings of 3A Certification Model for Power Terminal

SN	Authentication Parameters	Initial Setting	Remarks
1	5G terminal	10,000 devices	Appropriate reservation
2	Other access terminals	30,000 devices	
3	Concurrent authentication	80%	
4	Authentication cycle	One day	Adjust according to safety requirements
5	Key information	15 bytes	The number of bytes in the rule is uniform in the initial stage

Modeling for Certification System Application

The user authentication paradigm, which is based on 5G smart grid business scenarios and terminal kinds, is discussed in this section. The authentication model includes the following components, depending on the requirements of current business scenarios and the type of application protocol:

    i. Terminal classification: terminals are classified into two groups. The first type of terminal works with a 5G USIM card, while the second type connects to the 5G network.

    ii. Concurrent authentication parameters for terminals: the parameters of authentications that co-occur.

    iii. Authentication period: This relates to the time it takes to store authentication information.

    iv. Single authentication information length: the authentication rules limit the size of each essential authentication information item (in bytes).

PEI (IMEI), for example, takes up 15 bytes. The authentication model settings during the initial deployment are provided in Table 9.2, based on data format requirements.

5G Smart Grid Private Network Security Deployment Application

Due to regulatory constraints, power grid security is divided into production control and information management. Applications based on terminal security are enabled in establishing a 5G smart grid private network.

Edge computing nodes are built at critical industrial parks of the power grid for 5G power terminals and 5G wireless networks. In this context, power industry terminals and other terminals that support 5G connectivity are referred to as 5G industrial terminals. General industry terminals and dual-domain security terminals can be categorized based on the varying security needs of consumers. Dual-domain security terminals can give a higher level of data security and meet higher levels of protection needs than standard terminals. A tiny indoor 5G base station and an outdoor 5G macro base station provide seamless 5G wireless coverage throughout the park. Existing mobile and fixed networks in a specific area, such as Wi-Fi and optical fiber networks, are integrated to provide completely connected wireless coverage. All types of terminals can be operated this way.

Mobile Edge Computing (MEC) provides computational work offloading, edge access capabilities, and wireless access.

MEC servers' processing and storage capabilities can be used to deliver computing services to terminals in the park's working area.

Interaction delay can be minimized in this manner. Furthermore, the security level can be increased because the data generated by terminals are processed within the industrial park. Early-stage edge computing uses a built-in NFV mode, ensuring that the hardware meets the minimum requirements. The hardware required can be determined using the business scenario model in Table 9.3

TABLE 9.3   Main Hardware Requirements for Power 3A
System Construction

Hardware Configuration	Main Configuration	Main Function
Computing Unit	2*10 Core CPU	Add configuration to the general server
Memory	32G RAM 32 pieces	
Storage	3.2TB SAS SSD 8 pieces	

Terminal management, which includes security management capabilities such as unified authentication of user accounts, compliance content monitoring, and hierarchical content filtering, is at the heart of the security risk prevention and control platform. Control data is used to build secure large data applications. Packets are intercepted and deconstructed after the network deployment is complete for the secondary authentication process.

CHAPTER 10

5G Threat Surface and Threat Mitigation Control

The security capabilities and baseline recommendations of 5G architecture and protocols have substantially improved compared to earlier generations. The 3GPP(3rd Generation Partnership Project) standard includes enhancements for encryption, mutual authentication, integrity protection, privacy, availability, advancements for encryption, mutual authentication, integrity protection, and privacy, all of which are based on established 4G security processes. As carriers transition to 5G, attackers may look to target mobile devices and infrastructure using existing and new 2G, 3G, and 4G threats. Because of the large number of devices and the complexity of existing and newly deployed operator infrastructures, threats can arise from anywhere. Remember that not all threats can be addressed at a standard level. Appropriate resilience and mitigation mechanisms should be in place to account for hazards such as jamming, physical disruption, or Distributed Denial of Service (DDoS).

NETWORK THREATS

To make the transition to 5G secure, it's vital to consider some of the most common network risks that plague today's 4G networks.

Figure 10.1 depicts danger vectors that network operators examine today in the Core, Radio Acess Network (RAN), User Equipment (UE),

DOI: 10.1201/9781003264408-10

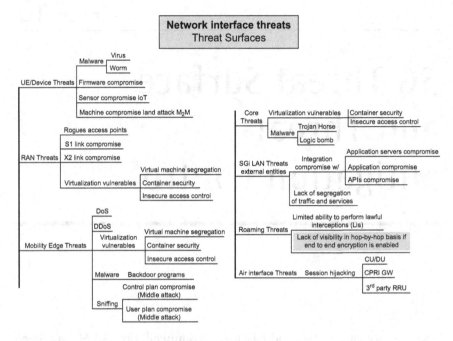

FIGURE 10.1   5G threat surface overview.

and other domains and how they use the concepts of visibility, segmentation, and controls to mitigate or remove such threats. These are traditional danger vectors or specific sources of entrance or weakness in the network. This isn't a complete list, but it serves as a starting point for examining what's new in 5G (for example, network slicing), the incremental threat surface of 5G, and the use cases that could lead to threats.

Some of the threats described in Figure 10.1 apply to 5G. Machine-to-Machine (M2M) may, for example, be constrained in power, processing, and memory resources in the UE/Devices threats vector, making it a poor candidate for providing security. M2M and other Internet of Things (IoT) use cases will necessitate network security, but not necessarily on the sensor or UE. Under-protected M2M communication can interrupt vital infrastructure if it is not adequately secured. 5G networks may support many linked entities and significantly increase bandwidth and real connections, which creates a threat landscape that can substantially impact network infrastructure.

Furthermore, network slicing allows for the division of mobile network operations, allowing the operator to be more flexible in applying security policies to each use case. Like any other network topology, network slicing will have its threat surface and necessitate best practices to ensure that the network is safely split. Because ultra-low latency use cases (such as those involving IoT and M2M) would require a broadly distributed network and security operations deployment to the Mobile Edge Computing (MEC) edge, segmentation is especially crucial in 5G. As a result, the edge and the facilitating network beneath it must be appropriately secured.

The various circumstances in which edge concepts can be used are enormous (mobile edge threat vector). There is no global perimeter to the edge that has been agreed upon. A fundamental network function like UPF can be put at the edge, but it may not be subjected to the same level of security scrutiny as if installed in a centralized location. Because of the open nature of mobile edge, malevolent adversaries can deploy their own devices and applications. This threat has the same effect as a Man-in-the-Middle attack, allowing someone to sniff traffic without permission.

Attacks between visited and home networks in roaming traffic (roaming threats vector) can occur. Because the security parameter to decode and extract the service content resides in the visiting network, end-to-end encryption limits the capacity to undertake lawful interceptions on inbound roamers. Furthermore, determining whether sufficient security countermeasures on a hop-by-hop basis might be problematic.

Threats arising from the SGi interface supporting various traffic and services can be problematic (SGi threats vector). The PGW (Packet Data Network Gateway) is connected to an external network via the SGi interface. The SGi interface could potentially support a variety of device types. A hack into a kind of gadget can have ramifications for other types of devices. These threats can target one or both devices (for example, a UE breach via a botnet attack) and/or mobile network infrastructure (for example, PGW User Plane DDoS), attacking one and potentially impacting the other.

IoT THREAT SURFACE WITH 5G

The following are four basic ideas to keep in mind when it comes to safeguarding IoT infrastructure:

1. IoT security should not be a last-minute consideration. IoT security should be addressed throughout the design phase rather than post-deployment.

2. Whether in healthcare, automotive, or energy, IoT requires several levels of security, including hardware, software, in-transit data, storage, network, application, and so on. The relevance of these levels and

how they interact are very contextual. This aspect must be considered in the overall architecture of IoT security.

3. The security of the IoT can only be as robust as its weakest link. The safety of a mobile phone is sometimes emphasized without regard for what occurs within the sprinkler control or car key programs that exist on it.

4. The most challenging IoT settings to secure are those with complex IoT devices (for example, industrial equipment and linked autos). For example, the implications of a hacked attached car can be far worse than those of a hacked connected energy meter or refrigerator.

5G allows hackers to enter networks across a larger area, including but not limited to mobile edge assaults. Furthermore, computing systems in-home or workplace settings might become a focus attack target, from IoT-enabled home devices to PCs at the edge and the data center or cloud. Without effective security solutions, combating the attack will be more difficult due to the high traffic volume generated by complex, integrated attacks.

Take a brief look at the vast landscape of IoT security (Figure 10.2).

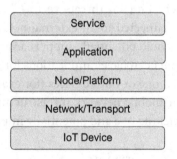

FIGURE 10.2   IoT security levels.

- IoT Devices—Many IoT devices will most likely be in exposed and vulnerable situations, and sensitive data stored on the devices may be tampered with. Malicious firmware and operating system updates are big issues.

- Network/Transport—Network connectivity allows devices or apps to interact securely with serving network nodes. Secure identification/authentication (credentials) and data transmission are required to secure this interaction. IoT network connectivity must be able to manage billions of devices while using a variety of access mechanisms and capillary networks.

- Node/Platform—Data and control commands must be secure on IoT platforms. Platforms are also in charge of ensuring that devices, users, third-party apps, and platform-based services are all isolated. Privacy concerns are one of the main impediments to adoption.

- Application—An application is a collection of microservices that work together to provide a service. These apps can be statically located or dynamically relocated to the best environment to run. The application code and the Platform it runs on will determine the security of the applications. When apps can migrate between platforms, they must do so safely.

- Service—The IoT provides a plethora of new services. Connected automobiles are a vital new service in which IoT will play a significant role and where ensuring security is critical. Safety will remain a top consideration for significant groupings of connected vehicles operating at high speeds. Backup measures the service can fall back on are required if network connectivity is lost due to failure or jamming. IoT can enable various other sensor-based services with varying degrees of relevance. The approach to securing specific IoT services will have to consider their uniqueness and the service's impact.

UE Threats

Smartphone adoption, a diversity of device form factors, more excellent data rates, a wide range of connectivity options (for example, Wi-Fi, Bluetooth, 2G/3G/4G), and the prevalence of open-source architecture are all aspects that make the UE a prime target for 5G network attacks. The various UE-targeted attacks in 5G networks can be divided into four categories:

1. Mobile to Infrastructure: DDoS attacks against 5G infrastructure are launched by a mobile botnet of many infected devices managed by an attacker's Command and Control (C&C) servers to render 5G network functions and services unavailable.

2. Mobile to the Internet: Using the 5G network, a mobile botnet with many infected devices controlled by C&C servers launches DDoS attacks on public websites.

3. Mobile to Mobile: Several infected devices conduct assaults on other mobile users to disrupt service or propagate malware (for example, viruses, worms, and rootkits).

4. Internet to Mobile: Malware contained in programs, games, or video players from untrusted app stores is sent to each UE by a rogue server on the Internet. Once downloaded and installed, the virus allows the attacker to access the device's stored personal data, spread the infection to other devices, and take control of the device to conduct attacks on different devices and networks.

RAN Threats

Because 5G will allow a variety of access networks, including 2G, 3G, 4G, and Wi-Fi, it will most certainly inherit those networks' security issues. The primary vulnerabilities and risks related to the RAN are described in this section.

A massive study has uncovered numerous security and privacy concerns in 4G mobile networks in recent years. Most 4G RAN layer attacks use Rogue Base Stations (RBSs) or IMSI (International Mobile Subscriber Identity) catchers to target IMSIs during the UE's initial network connection operation or paging attacks employing the IMSI paging feature. In such assaults, information about specific IMSIs may be collected and later exploited in other forms of attacks. Fortunately, all access types, including licensed RAN and unlicensed Wi-Fi, are likely to be addressed by 5G technology and protocols, which are expected to address identified concerns at this layer. An unencrypted IMSI, for example, will not be transmitted over 5G.

Multiple-Input Multiple-Output (MIMO) antenna arrays and beamforming will be used in 5G devices and networks. Many 5G systems will use a millimeter wave (mmWave) spectrum and other frequencies. mmWave is believed to be as secure as other spectrum bands. Data and signals sent and received at the radio layer should be securely secured and protected at higher layers.

RBS Threat

The RBS threat is one of the threats that different mobile networks face, including 5G. To assist a Man-in-the-Middle (MiTM) attack between the mobile UE and the mobile web, the RBS poses as a legal base station. An attacker might theoretically use an RBS to launch various mobile users and network attacks. An RBS attack usually requires one or more software-defined radios that do not have access to or membership in an operator's core network, restricting the attack's potential impact to early communication phases before UE authentication. Theft of user information, altering with transmitted information, tracking users, compromising user privacy, and generating a DoS attack for 5G services are all possible outcomes of these attacks. However, practical attack routes remain disputed and possibly limited due to the limited range/coverage of realistic, portable software-defined radios, conventional physical security limits, and other variables. Other industry organizations are conducting extensive analyses of this issue, which would surpass the space provided in this whitepaper.

The RBS threat has been around since GSM networks and has continued to evolve and survive as mobile networks have grown. Compared to 4G and legacy networks, 5G networks will bring significant security improvements. Despite these security improvements, RBS-based malware may still target 5G networks. As examples, consider the following threat vectors:

- An attacker can use the 5G/Long Term Evolution (LTE) interworking requirement to perform a downgrade attack

- A hacked 5G tiny cell poses an RBS risk to 5G networks and customers

- In idle mode, an attacker can use a lack of gNB authentication to force a user to camp on an RBS, resulting in a DoS (such as public safety warnings, incoming emergency calls, real-time application server push services, etc.)

Subscriber Privacy Threats

The cellphone business has always placed a premium on subscriber privacy. Because of the media and regulators' increased focus as the 5G era begins, subscriber privacy has become even more critical. For example, there have been various news stories about allegations of mass spying. Unknown RBSs have also been reported tracking people in big cities and engaging in questionable behavior.

Core Network Threats

Because of their IP-based service design, 5G networks may be subject to standard Internet IP attacks, such as DDoS attacks. Many compromised mobile devices can also launch both user plane and signaling plane assaults against 5G core network services, thanks to malicious C&C servers. As a result, critical services are degraded, and legitimate users cannot access them.

The critical network functions of 5G are the Access and Mobility Management Function (AMF), Authentication Server Function (AUSF), and Unified Data Management (UDM). UE authentication, authorization, and mobility management are all provided by the AMF. The AUSF keeps data for UE authentication, while the UDM maintains data for UE subscriptions. These tasks are crucial in 5G; a DDoS assault from the Internet or a mobile botnet against these functions might drastically impair the availability of 5G services or perhaps cause network disruptions.

For non-3GPP access, 3GPP advises employing Internet Protocol Security (IPSec) encryption. An attacker can use many infected mobile devices to execute a DDoS assault on 5G core network services by exploiting many IPSec tunnel formation requests.

Network Functions Virtualization (NFV) and SDN Threat

To properly support the increased levels of performance and flexibility necessary for 5G networks, new networking paradigms such as NFV and SDN must be used. These new techniques, on the other hand, pose unique dangers. The integrity of Virtual NFs (VNFs) and the security of their data may depend on the isolation properties when using NFV. They will also be reliant on the entire cloud software stack in general. In the past, vulnerabilities in such software components have frequently surfaced. Ensuring a thoroughly reliable and secure NFV environment remains a significant difficulty.

Additionally upgraded transport networks and improved networking ideas, such as SDN, are projected to connect the 5G cloud data centers. SDN centers the risk of control applications interacting with a central network controller incorrectly or maliciously, thus causing widespread disorder. SDN separates forwarding and control, resulting in an interface between the SDN controller and the SDN switch. The total system becomes more vulnerable as a result of this interface. It could enable attacks on the controller-switch communication's integrity, secrecy, and DoS assaults.

Furthermore, it may enable attacks targeted at taking control of switches and controllers by exploiting flaws in protocol software or interface settings. However, protecting such an interface is a well-known task with widely available solutions, such as using IPsec or TLS to cryptographically safeguard lawful connections while excluding communication from harmful third parties.

Interworking and Roaming Threats

Compared to 4G, roaming in 5G uses new protocols that provide greater flexibility and unknown threats. The embedded security at 5G roaming links is addressed in the following roaming concerns in the 5G architecture.

- The Security Edge Protection Proxy (SEPP) node has been added to the 5G architecture to terminate signaling communications between PLMNs (Public Land Mobile Network) over inter-exchange/roaming links.

- The interconnection model will be comparable to today's 3G and 4G networks' SS7 or DIAMETER connections. However, on inter-exchange/roaming links, the application layer protocol like HTTP/2 will enable encryption.

- The incorporated application layer encryption at the SEPP will defend against the SS7 and DIAMETER protocols' known inter-exchange/roaming weaknesses.

Threats in Network Slicing

Improper isolation between network slices (Inter-Slice Isolation) and improper isolation between the components of the same piece are two of the most common reasons for network slicing hazards (Intra-Slice Isolation).

Danger could be moved between the slices if one of the devices in the IoT slice is infected by malware that exploits a vulnerability in the IoT device. Critical slices (such as the V2X slice) will be impacted as a result.

The attack might be multi-factored by allowing the virus to drain the slice's resources, inflicting DoS to the actual subscriber. An attacker could also deplete resources shared by many slices, resulting in service degradation or denial in other slices. As a result, the network services available suffer a significant degradation.

5G Core (5GC) is a cloud-native architecture with all functions virtualized to provide the increased flexibility needed for network slicing. However, this opens the door to a new threat. Data exfiltration is caused by side-channel attacks combined with a lack of isolation between network slices. This is crucial in mobile network-sensitive layers like billing, charging, and subscriber authentication.

Suppose the slices and components within the slice are not adequately segregated. In that case, the attacker could access other slice components via the infected device or endpoint in another slice. The compromised device allows the attacker access to the slice resources. The additional slices are eventually exposed, and data is exfiltrated to an external server (a C&C center, for example). Once the attacker has gathered all of the network's information inside the firewall, they could use it to launch an assault on subscribers. Furthermore, the attacker could use the information to make dishonest financial advantages.

Customers can get personalized services thanks to network slicing. It is conceivable for 5G systems to provide standardized APIs to build, edit, delete, monitor, and update network slice services based on operator policies. If not secured, slice management also contains significant threat vectors. The management interface between the Network Slice Management Function (NSMF) and the Communication Service Management Function (CSMF) or between Communication Service Provider (CSP) and Communication Service Customer (CSC) is also specified according to 3GPP standards specifications. Interfaces are also defined for a Network Slice Instance (NSI) management phase, supervision, and performance reporting during the operation phase.

If not secured, attackers could access these network slice management interfaces. With this access, attackers can build network slice instances that use many network resources or a lot of network slice instances. As a result, network resources are depleted, resulting in DDoS assaults. Attackers could instigate fraudulent activities such as bogus charging by replaying management messages. An attacker might also listen in on the transfer of supervision and reporting data and collect sensitive data from launching attacks against active network slice instances.

Mitigating Threats in Network Slicing

High-security layers are required for network slicing design, which allows several logical networks to operate as essentially separate business activities on a shared physical infrastructure—isolation between slices and isolation inside the components of the slice, in particular. In the event of a malicious attack, this prevents vulnerabilities from spreading to other components inside and between slices.

We may differentiate two types of isolation in Network Slicing: resource isolation and security isolation. The former refers to the fact that a network slice's resources (such as processing, storage, and networking) cannot be "hijacked" by another slice. While a slice is the subject of a DoS attack, resource isolation ensures that the slice's required resources stay available even when other slices try to scale out and acquire additional resources.

Information in one slice cannot be viewed or modified by other slices sharing the same infrastructure as security isolation.

Virtualization layers, such as a hypervisor, provide isolation as an essential feature. As a result, good NFV security—including a solid virtualization layer and overall cloud stack implementation—can enable both types of isolation in NFV systems. To ensure isolation, dedicated virtual networks per slice can be created for transport between different hardware platforms, including distributed cloud deployments. At the same time, SDN allows for highly dynamic and flexible control of various virtual transport networks sharing the same transport infrastructure.

Non-virtualized equipment can also be shared between slices. For example, base station equipment is in charge of the lower protocol levels. Equipment-specific methods, in this case, must ensure isolation. When numerous slices share a single cell, a standard radio scheduler can be used to implement scheduling policies. This ensures that each slice in that cell receives radio resources through the slice's service-level agreement (SLA).

The role of each slice as an isolated logical network or network portion necessitates the use of cutting-edge network security mechanisms. This could include perimeter security and network zoning via firewalls (usually virtual firewalls rather than physical firewalls), traffic separation, intrusion and anomaly detection, cryptographic traffic protection, etc. Slices enable security measures to be tailored to the specific requirements of the application(s) supported by that slice. An IoT slice, for example, can have a different set of authentication and encryption algorithms than an eMBB slice.

To ensure intra- and inter-slice security, a quarantine slice should be implemented. Thanks to segment routing and segmentation technologies, the quarantine slice can be implemented in transport and the data center. This design combines these two segmentation methods to create a segmented network that provides visibility and agility in threat detection and management from end to end.

Software-Defined segmentation is achieved by combining segment routing and network-wide policy enforcement. Access policies for users, apps, and devices can be enforced using software-defined segmentation applied to IoT devices, M2M-based devices, and enterprise network devices.

Segment routing is used by software-defined segmentation in the data center element of the 5G architecture. When software-defined segmentation is utilized, an identification and segmentation policy controller can define and maintain security group tags. The controller can also share group data with other group-based policy schemes, allowing traffic segmentation and communication restrictions.

The 5G Americas White Paper described network interfaces in The Evolution of Security in 5G. This security strategy replaces the reliance on extensive lists of IP addresses with a flexible, automated model that is easier to administer and more effective against new and evolving threat vectors. Namespaces, ACLs (access control list), TAGs, and the software-defined segmentation controller can all be used to create slice isolation. In addition, applications that capture telemetry and give secure outcomes can provide visibility on the slices. Some of the safe effects are: Tokens are used to secure the core VNF components; a Web Application Firewall (WAF) is used to secure the API interfaces, and an allow listing application behavioral approach is used to enforce application policies.

Understanding the normal behavior of an IoT device might help you spot when it's acting strangely. By identifying anomalies or changes in the conduct of IoT devices, systems can discover areas of concern. Anything from a software upgrade to criminal activities could cause the abnormality.

Finally, encrypted traffic is a hurdle in monitoring IoT devices. Although vendor-specific approaches address encrypted traffic's security challenges, the industry requires a more advanced strategy than today's man-in-the-middle techniques. Enabling virus detection in encrypted communications is crucial for a better understanding of traffic behavior and conducting more in-depth research.

One of the most critical security components is attack prevention. Today's technologies may help by providing security from the cloud, blocking dangerous destinations before a connection can be established. A DNS-based security technique based on name resolution identifies Internet communications from every device on the network, blocks malicious domains, and destroys C&C callbacks.

5G THREAT MITIGATION CONTROLS: IoT, DDoS ATTACKS, AND NETWORK SLICING

The capacity of a network to automatically build and execute numerous logical networks as essentially separate business operations on a common physical infrastructure is known as network slicing. Slicing is predicted to be an essential architecture component of the 5G network, fulfilling the majority of the 5G use cases, although being used sparingly today for Enterprise use cases.

The network slice, according to 3GPP specifications, is a complete logical network (offering Telecommunication services and network capabilities) that includes the Access Network (AN) and the Core Network (CN). Thanks to network slicing, multiple logical networks can be managed as essentially independent business processes on a single physical infrastructure. In practice, this translates to the idea that the mobile network could be divided into a series of virtual resources. Each one is referred to as a "slice," and can be used for many purposes. A slice could be assigned to a mobile virtual network operator (MVNO), an enterprise customer, an IoT domain, or any other group of services that makes sense (for example, mobility as a service). The access point name (APN) idea utilized in today's mobile network is extended by a network slice.

5G Network Threat Mitigation

Separating a network with a significant danger surface, such as 5G, into parts and examining the threats and mitigations for each piece is one strategy to address threat mitigation.

For threat mitigation, you cannot ignore the importance of endpoint security (anti-malware, day 0, and day 1 protection on the endpoint). This protects not only the UE (for example, a phone or an iPad) but also the RAN by preventing the formation of a botnet that would attack the RAN. Commonly used DNS protection measures impede attacks at the first stage of the malware kill chain by preventing C&C contacts with known "bad actors." This is only one example; operators will have their use cases that expand.

All forms of security require visibility. Visibility gives you a real-time snapshot of how the network is doing. Threat feeds enable that image to be used in the face of both known and unknown threats. We employ policy and segmentation to ensure we know what is odd or strange. The network is divided to prevent threats from spreading and impacting other functions or workloads. A web application firewall, API protection (commonly referred to as a cloud service access broker type function), and malware protection are all used at this point in the network to mitigate DDoS threats (volumetric and application-based), web application threats via a web application firewall, and malware protection. These controls ensure that the "edge" is protected.

5G introduces orchestration layers, NFV, containers, microservices, and virtualized implementation of essential advanced packet core functions.

Slicing is also included in MEC technologies. For example, at the edge, there may be numerous tenants for V2X services (MEC App1, MEC App2, and so on), each running on a particular slice. Over time, these tenants may need to communicate with one another and with neighboring MEC instances.

Threats to the distributed core by interface include, but are not limited to, the following:

- N2: On NG2, sniffing over untrusted networks creates a vulnerable control plane.

- N4 (between centralized and remote data centers): DoS attacks on the NG 4 interface from compromised remote data centers using an untrusted network for the NG 4

- Compromised MEC servers in remote data centers

- N6 (Internet-facing): NG 6 interfaces with low latency exposed to the Internet and vulnerable to DDoS and DoS threats from the Internet.

5G bring issues of distributed deployment, orchestration, and scale-up/scale-out with automation to stay up with threats on a distributed core architecture.

In the context of the 5G design, risks at the virtualization layer must also be addressed. In important NFs, today's networks are highly virtualized. The operator's backbone network connects several smaller data centers that are spread out around the country to a few larger data centers. This infrastructure necessitates visibility into application dependencies and traffic patterns, which is fed into the broader analytics function. On top of such infrastructure, an orchestrated NFVi layer exists. 5G accelerates the shift to highly virtualized workloads and, in some circumstances, the cloud migration of crucial network components (CUPS model for Control and User Plane Separation). CUPS is not a 5G feature per se, but it is an example of the increased trust boundaries and threat surface that 5G network deployments have created. The increasing number of application policy enforcement instances must also account for slicing in the distributed core mitigation. Virtualized workloads introduce a new set of dangers, including, but not limited to:

- Trust and compromised VNFs.

- Sprawl and VM hopping

- Microservices that have been compromised

- Vulnerabilities in container images

Visibility, segmentation, DNS level security (for example, known bad talkers, bad domains), and detection of aberrant flows all contribute to the virtualization of 5G architecture's core layer of security.

IoT and DDoS Threat Mitigation

With 5G technology, there will be numerous ways to minimize IoT dangers and threat surfaces, and these strategies are discussed in this section.

IoT Device

Encryption and integrity protection are required for sensitive data stored in non-secure physical device locations. Even in the event of malware infestation, devices must cryptographically check firmware and software packages during system startup or update and maintain the ability to receive remote firmware upgrades. In an update failure, sufficient storage must be available for automatic rollback. However, malicious rollbacks to older software/firmware versions must be avoided because they reintroduce old vulnerabilities.

It's vital to have security isolation between device-resident programs. One solution is to implement hardware-based application isolation. This entails a 'root-of-trust' method to avoid a compromised operating system. Although specialized hardware is typically used to provide this functionality, Trusted Execution Environments (TEE) can also be used. In standard processors, the TEE is separated from the client-side execution environment and referred to as the Rich Execution Environment (REE). The use of TEE is preferred for low-cost devices. Global Platform makes the TEE specification set available to the public.

Today, cryptographic methods, especially asymmetric algorithms, are substantially faster and better suited for IoT than traditional techniques. For at least some circumstances, lightweight cryptography may be appropriate. Protection against side-channel attacks is critical for IoT devices that operate in open areas since it prevents the leaking of keying material via timing information, electromagnetic signatures, power usage, and other means.

Network/Transport

Mobile operators may take advantage of their unique position as connection and platform providers in the IoT industry. LTE Category M-1 (LTE-M) and Narrow Band-IoT (NB-IoT) technologies are preferable to unlicensed access to provide more stable worldwide connectivity. Mobile networks can improve IoT security by enabling device administration, secure boot-strapping, and confirming device location or platform trustworthiness.

Device credentials are typically pre-provisioned on detachable UICCs (Universal Integrated Circuit Card). Remote provisioning and credential management are possible via an embedded UICC (eUICC). By actually generating credentials on the device, security breaches are lowered. The usage of a TEE that is already embedded in the baseband processor is a reasonable next step. This combination provides benefits such as lower hardware costs and power consumption, increased speed, and the ability to securely modify credentials and network configuration.

The IoT affects a wide range of ecosystems. For some use situations, the ability to securely bootstrap connection credentials from the device and/ or application credentials from connectivity is critical. Consider a cus-tomer who wants to work with a single connectivity aggregator on a single service layer agreement. The number of slices a device can use is another essential part of provisioning and administration that could benefit from this flexibility.

Node/Platform

IoT systems can and should manage the lifecycle of IoT devices from installation through decommissioning, ideally with as little user involve-ment as possible.

An IoT device will often automatically bootstrap itself into active ser-vice during the device installation step, utilizing pre-configured creden-tials (keys IDs) stored in a secure hardware module or baseband processor. The respective IoT platform will perform initial configuration processes, including firmware updates, application configuration, and credential provisioning for application-layer services.

The Platform should implement security standards such as permis-sion and access control during device operation and any required delta updates in software, credentials, storage, and so on. The Platform must be able to remotely remove all sensitive data stored on the device when it is decommissioned.

Application

IoT applications should be deployed on secure platforms leveraging cloud infrastructure with roots of trust. Lightweight IETF security protocols, such as an authorization framework based on OAuth (IETF) suitable for limited situations, can protect data exchange between IoT apps or between applications and devices.

IPsec and TLS alone may not be adequate to protect against intermediaries. These protocols support only trust models that can guarantee completely trustworthy endpoints. Access to information should only be granted to those who have a legitimate need to know. End-to-end security at the application layer is required to achieve this goal. Information containers at the application level, rather than at lower tiers in the protocol stack, are the recommended method for securing message exchanges. Confidentiality, integrity, and origin authentication are all possible with these containers.

Service

The current connected vehicle scenario discussed earlier is used as an example to demonstrate service-level security. Thousands of sensors, actuators, and a code base are dispersed over thousands of embedded processors in connected vehicles. Isolation, both mental and physical, is crucial in this situation. For example, a breach in the entertainment system must not impact the steering system. Firmware changes must ensure that associated subsystems are compatible. Almost all accidents may be avoided with vehicle-to-vehicle communication. Accidents caused by malfunctioning machines may never be removed, but establishing secure communication has the potential to make transportation much safer.

There are numerous other scenarios in which the IoT might improve public safety. By putting sensors and cameras into traffic lights, for example, motorists could be made aware of pedestrians in advance. Another area where IoT can have a substantial positive impact is emergency response. Free traffic lanes for emergency vehicles could be formed automatically, missing children could be found or tracked more efficiently, and natural and man-made calamities could be better monitored and contained. Because of the critical nature of these circumstances, service-wide security is required to prevent misuse or even suspicion of abuse. Of course, public safety must always be balanced against the need for privacy (for example, the right to be forgotten). To achieve that balance, a secure IoT service infrastructure can be tuned.

Security Requirements For 5G Network Massive IoT (MIoT) Threats

To prevent 5G service disruption caused by MIoT botnets deployed for DDoS RAN attacks and to provide 5G service resiliency, deliberate security standards for the 5G network are required. The foundations of these security needs are the detection and mitigation of DDoS attacks against the 5G RAN, also known as 5G RAN overload functions. Collaboration between the 5G standards community, 5G operators, and 5G RAN manufacturers will be required to meet these security requirements. Although each operator's unique 5G network implementation may offer some security against this type of assault, it will not be sufficient. In order to detect and mitigate these types of assaults in real time, 5G RAN components will need to play a substantial role. The 5G standards community and 5G RAN suppliers will be crucial in this regard.

Detection of DDoS Attacks Against the 5G RAN

Detailed aspects of the attack must be investigated to detect a DDoS attack by MIoT botnets on an operator's 5G RAN. The following excerpts are from the previously outlined attack scenario: Malicious hackers command their MIoT botnet army to simultaneously reboot all devices in a specific or targeted 5G coverage region. Excessive malicious attach requests will result, resulting in a harmful signaling storm. The detection requirements can be formulated using these details.

Given the required real-time response, the 5G RAN components directly touched by this type of attack will be the most effective parts to play a vital role in the detection process. The Radio Unit (RU), Distributed Unit (DU), and Centralized Unit (CU) are the components of the 5GRAN NR or gNodeB. Given the functions of these components, the Central Unit Control Plane (CU-CP) will be the best component to use for detecting this type of attack.

The CU-CP is the most efficient site for embedding detection functions because it is responsible for maintaining the Radio Resource Control (RRC) links. An adjustable threshold for all aspects of RRC connection requests, analytics algorithms to determine if it is a DDoS event (based on threshold), volumetric anomaly, timing, Radio Network Temporary Identifiers, and so on are the key software elements of the detection functions that must be embedded in the CU-CP. Through open interfaces, the adjustable threshold function and analytics function should be able to receive updates from an external Machine Learning (ML) and Artificial Intelligence (AI) platform.

Mitigation of DDoS Attacks Against the 5G RAN

For the mitigation of a DDoS attack on a 5G RAN, the same attack scenario will be considered. Once the CU-CP recognizes the DDoS attack, some mitigation step is required. The CU-CP would be the most effective 5G RAN component to counter this attack. This is because the CU-CP is critical for managing RRC connections, making it perfect for blocking malicious Attach Requests in large numbers. The described procedures for identifying and mitigating this threat will show inherently closed-loop automation.

Protecting 5G Networks Against DDoS and Zero-Day Attacks

Both the control and data planes of 5G networks are vulnerable to attacks. Threats to the control and data planes, as well as mitigation techniques, are outlined below.

The control plane is the first example. A series of communications must be exchanged between the evolved NodeB (eNB), next-generation NodeB (gNB), and eventually the MME before the UE may create a connection (for example, to make calls). A signaling storm could occur if an attacker can seize control of numerous devices and force them to rejoin (for example, by restarting them). Compared to LTE networks, 5G networks can support 100× more devices and 1000× more capacity per unit area.

Another example is an attacker who uses legitimate devices on an operator's network to launch a denial of service attack against the operator or a third party. At the data plane level, such assaults generate considerable traffic.

Although these attacks occur on the data and control planes, they are fundamentally similar. Network devices produce anomalous amounts of traffic (of various types) in both circumstances, and the traffic is defined by some similar, albeit intricate, characteristics.

These attacks can be detected in a variety of ways. Supervised models perform exceptionally well in network intrusion detection when given good training data. On the KDD99 dataset, which is frequently used in ML research and intrusion detection systems, simple deep neural networks (DNN) do extraordinarily well in detecting attacks. Creating good labels remains difficult, which can be solved via an unsupervised process.

It is suggested to use a combination of these two ways. The initial step is to extract statistics from 24-hour sliding windows and feed them into an anomaly detection system. There are numerous viable options here. Approaches based on the Mahalanobis distance function, auto-encoders, and isolation forests work well.

This method will result in a large number of false positives. Reduce false positives by recognizing that DoS service attacks establish connections with shared commonalities. At this process level, simple vertical characteristics (counting the number of abnormal connections per gNB, or a specific User-Agent string or Type Allocation Code) can be utilized to develop rudimentary algorithms to limit false positives. One of the methods is to use a clustering algorithm like K-Nearest Neighbors to identify groupings automatically. Representing the data that can be fed into a Convolutional Neural Network (CNN) for anomaly identification is a more robust approach.

Network Slicing Security Threat Mitigation

There are several methods for attaining security isolation, each with its own set of advantages and disadvantages. Depending on the threat model, the problem of secure isolation can be framed in two ways. Isolation can be achieved in one way by running untrusted programs within a security perimeter.

On the other hand, hardening a system will protect the execution of trusted but susceptible programs with a larger attack surface. For example, Internet-facing applications (Web Servers, Email Servers, and DNS) are trusted, yet they must be protected to prevent vulnerabilities from being exploited.

CHAPTER 11

Role of AI in Mitigation of 5G Attacks

Cyber-attacks are widely regarded as one of the most dangerous threats to global security. The attack vectors are not the same as they were five years ago in terms of availability and efficiency. Improved technology and more effective offensive strategies enable cybercriminals to launch cyberattacks on a large scale with a more significant impact. Intruders use new ways to breach systems and launch increasingly complete Artificial Intelligence (AI)-based plans. Similarly, businesses have begun to employ sophisticated defense systems based on AI to combat AI-powered cyber-attacks.

Security experts have spent a lot of time figuring out how to leverage AI to exploit its capabilities and incorporate them into technological solutions. AI provides defense systems and services to detect and respond to cyber threats. The application of AI to security has proven to be advantageous. According to many IT specialists, the key motivation for AI adoption in organizations is security. AI not only improves overall cyber security but also automates detection and mitigation processes.

For threat detection, businesses employ security information event management (SIEM) to collect a vast amount of data from multiple sources. It is impractical for a user to sift through such data in order to find potential flaws. Furthermore, AI aids in the detection of anomalies in both technology and user operations. AI-based algorithms effectively scan the system and analyze diverse information sources to find vulnerabilities. Anomaly

DOI: 10.1201/9781003264408-11

detection is one area where AI may enhance a company's security defense. Looking at previous instances, it also discovers various functions for preventing attacks (Machine Learning [ML]).

Software-based cyber-attacks will be the most common type of cyber-attack on 5G. Software protections must also be used to combat them. AI-based defense solutions are the most effective defense against these virtual attacks. Once installed, AI-powered software security products will continue to improve and adapt to an ever-changing environment, increasing a network's defense levels through self-learning.

However, we cannot rely on defense products added retrospectively. Security should be considered at every level of the network development lifecycle, not just included in a finished product, to design resilient cyber security solutions. In order to secure the network's integrity, security must be built into the hardware, firmware, and software development. In reality, regulatory agencies are likely to impose baseline security standards for all 5G hardware and software.

WHAT AI AND ML CAN DO FOR CYBER SECURITY

AI and cyber security have been hailed as game-changing technologies that are closer than we might imagine. However, this is only a part of reality, and it should be approached with caution. The reality is that we may have to settle for more incremental advances in the future. In comparison to a totally autonomous future, what may appear incremental is nevertheless leaps beyond what we've been capable of in the past.

It's crucial to contextualize the current pain points in cyber security as we investigate the potential security implications of ML and AI. We've long considered many processes and elements as usual that can be addressed using AI technologies.

HUMAN ERROR IN CONFIGURATION

Human mistake is a major source of cyber security flaws. Even with huge IT teams involved in setup, correct system configuration, for example, can be complicated to manage. Computer security has become more layered than ever before due to ongoing innovation. Responsive tools may be able to assist teams in identifying and resolving issues that arise as network systems are replaced, upgraded, or updated.

Consider how cloud computing and other contemporary Internet infrastructure can be layered on top of older local frameworks. In order to safeguard business systems, an IT team will need to verify interoperability. As they juggle countless updates with routine daily support work, teams become exhausted by manual processes for assessing configuration security. Teams could obtain immediate guidance on newly discovered difficulties thanks to sophisticated, adaptive automation. They could get advice on how to continue or even have mechanisms in place that modify settings automatically as appropriate.

HUMAN EFFICIENCY WITH REPEATED ACTIVITIES

Another issue in the cyber security sector is human efficiency. In a dynamic setting like ours, no manual method is perfectly reproducible every time. One of the most time-consuming jobs is setting up each of an organization's many endpoint machines individually. IT teams frequently return to the same workstations after initial setup to rectify misconfigurations or outdated settings that cannot be addressed via remote upgrades.

Furthermore, the scope of danger can swiftly vary when personnel is tasked with responding to it. A system based on AI and ML can move with a minimal delay where unforeseen hurdles slow the human focus.

THREAT ALERT FATIGUE

If not appropriately managed, threat alert weariness can become a new source of weakness for organizations. As the layers of security get more intricate and expansive, attack surfaces are expanding. Many security systems are programmed to emit a bombardment of merely reflexive alarms in response to various recognized vulnerabilities. As a result, human teams are left to sort through prospective decisions and take action due to these individual impulses.

Due to the high volume of warnings, this decision-making level is very stressful. Decision fatigue eventually becomes a daily occurrence for cybersecurity professionals. While taking proactive measures to address these risks and vulnerabilities is preferable, many teams lack the time and resources to do so.

Sometimes, teams must choose to address the most pressing issues first, allowing secondary goals to fall by the wayside. Using AI in cybersecurity initiatives can help IT teams manage more of these threats more effectively and practically. Automated labeling can make dealing with each of these dangers much easier. Aside from that, the ML method may be able to address some of your worries.

THREAT RESPONSE TIME

Threat reaction time is one of the most critical indicators of a cybersecurity team's effectiveness. Malicious attacks have been known to proceed fast from exploitation to deployment. Threat actors in the past had to trawl through network permissions and deactivate security across multiple networks for days or weeks before initiating an assault.

Unfortunately, cyber protection professionals aren't the only ones that benefit from technological advancements. Since then, cyber-attacks have become increasingly automated. Threats like the recent LockBit ransomware outbreaks have significantly sped up assault times. Currently, some attacks can take up to a half-hour to complete.

The human response can lag after the initial onslaught, even with known attack types. As a result, many teams have focused their efforts on responding to successful assaults rather than preventing attempted ones. On the other hand, Undiscovered attacks represent a distinct danger in and of themselves.

Thanks to ML-assisted security, data from an attack may be quickly sorted and readied for investigation. It can deliver more concise reports to cybersecurity teams to make processing and decision-making easier. This sort of protection can provide recommendations for limiting further damage and preventing future assaults and reporting.

NEW THREAT IDENTIFICATION AND PREDICTION

Another element that influences cyber-attack reaction times is discovering and predicting new threats. As previously stated, existing threats already have lag time. Unknown attack kinds, behaviors, and tools might also fool a team into taking too long to react. Worse, more subtle dangers, such as data theft, might go undetected for long periods of time. According to a Fugue poll conducted in April 2020, 84 percent of IT teams are afraid of their cloud-based systems being hacked without their knowledge.

The constant evolution of attacks, which leads to zero-day exploits, is always a concern for network protection operations. However, there is some good news: cyber attacks are rarely created from the start. ML has a pre-existing path to work from because it is frequently built on top of behaviors, frameworks, and source codes from previous attacks.

To help recognize an attack, ML-based programming can help highlight commonalities between the current danger and previously discovered threats. This is something that people are incapable of doing in a timely manner, emphasizing the importance of adaptive security mechanisms.

In this regard, ML may make it easier for teams to predict new risks and reduce lag time as a result of greater threat awareness.

STAFFING CAPACITY

Staffing capacity is one of the difficulties impacting many IT and cybersecurity organizations worldwide. The number of qualified specialists available to an organization may be limited depending on its demands.

However, the more typical situation is that hiring human help can eat up a significant portion of an organization's budget. Supporting human workers necessitates reimbursing them for their daily labor and assisting them with their ongoing education and certification needs. As a cybersecurity practitioner, staying current is complex, especially given the constant innovation that has been mentioned throughout this conversation.

With a smaller crew to manage and support it, AI-based security systems can take the lead. While this personnel will need to stay updated on cutting-edge AI and ML topics, the reduced staffing requirements will save money and time.

ADAPTABILITY

Adaptability is not as evident a problem as the other points stated, but it has the potential to alter an organization's security capabilities drastically. Human teams may be unable to tailor their skill sets to your specific needs.

You may find that your team's effectiveness is hampered due to a lack of training in specific methodologies, tools, and systems. Even seemingly straightforward demands like implementing new security procedures can take a long time with human-based teams. This is simply human nature; we cannot learn new methods of doing things instantaneously and must take time to do so. Appropriately trained algorithms can be molded into a bespoke solution exclusively for you with the correct datasets.

USE OF AI IN CYBER SECURITY

AI in cyber security is a subset of fields such as ML and deep learning cyber security, but it has a distinct function to play.

At its foundation, AI is focused on "success," with "precision" being given less weight. The ultimate goal in complex problem-solving is natural solutions. Actual autonomous decisions are made in a true AI implementation. Its programming is focused on finding the best solution in a given situation, rather than just the dataset's hard-logical conclusion.

To comprehend how modern AI and its underlying disciplines work today, it's essential to understand how they work. Autonomous systems, particularly in the sphere of cyber security, are not within the scope of widely deployable systems. Self-directed systems are what most people think of when they think of AI. On the other hand, AI technologies that support or supplement our protection services are viable and available.

The interpretation of patterns created by ML algorithms is the perfect function for AI in cyber security. Of course, modern AI isn't yet capable of interpreting results with the same skill level as a person. While effort is being made to advance this discipline in the search for humanlike frameworks, genuine AI remains a long way off, requiring machines to take abstract concepts and reinterpret them across scenarios. In other words, the AI rumors would have you believe that this level of creativity and critical thinking is very close.

Is AI Capable of Enhancing the Security of Cloud-Based Data?

AI can improve security across the tech stack, from cloud deployments to data-accessing endpoints. Rule-based systems may not be able to detect security vulnerabilities across the stack, necessitating the creation and maintenance of complicated rules over time. With AI, the thresholds are automatically set based on the data and seasonal trends.

At the cloud level, AI can limit access to privileged information and avoid various assaults such as Distributed Denial of Service (DDoS), zero-day exploits, etc.

Why Is AI Being Used in Cyber Security?

It is an obvious question why add another costly technology to a company's cybersecurity platform? Especially, in a department where many upper management employees consider to have a poor return on investment. The following are some reasons:

- Enterprise security and privacy standards have become a symbol of a company's reliability. Regardless of how competitive your service is, a security breach or lax privacy procedures can harm an organization's reputation and customers will flee to competitors.

- It's only fitting that you put up your best effort to keep on top of the cybersecurity game. By incorporating new technologies like AI into your security processes, you can show your consumers that you've been paying attention and that you're in it for the long haul.

Apart from preserving a positive public image, AI can assist an organization to stay ahead of cyber attackers. We all know that the pandemic world has democratized sensitive data access. Private networks and business devices are no longer the only places where confidential information can be viewed; they can now be accessible from anywhere on any device. It provides hackers with several possible entry points to fraudulently access your personal and confidential company data. By compromising access points, attackers employ advanced tools like AI to abuse unwary end-users and obtain access to privileged information.

Another drawback is that traditional (non-AI) security approaches rely on static thresholds to function. Attackers can take advantage of the system by avoiding static thresholds.

The moment has come to use AI to improve security and privacy protection. Some several real-world cyberattack scenarios and how AI could help cybercrime investigators.

1. Example: When the number of failed logins to access private information surpasses 10 per minute, a SIEM solution is programmed to warn the enterprise. Even if a brute-force attacker manages nine failed logins every minute, they will remain anonymous.

   Solution: Set elastic thresholds with little to no human interaction as a solution. In addition, AI may track login patterns and set thresholds based on a variety of factors such as the time of day, the day of the week, and other recent trends in information access. For example, different criteria may be required on a Monday at 9 a.m. and a Saturday at 3 a.m.

2. For example, an incorrectly adjusted threshold could cause alarm fatigue in the person in charge of monitoring SIEM system warnings.

   Solution: AI can reduce alert fatigue by detecting common, rare, and previously unknown patterns and adjusting the alert priority accordingly.

3. For example, monitoring access to every potential ransomware and phishing website is nearly impossible for cybersecurity personnel.

   Solution: AI may be deployed at endpoints to help identify and quarantine harmful websites, allowing for improved data-access practices combined with multifactor authentication and zero-trust security measures.

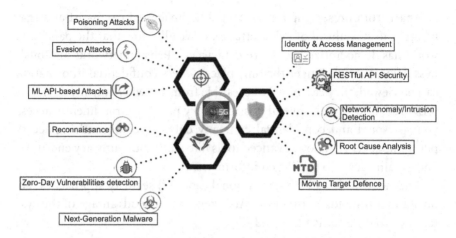

FIGURE 11.1 AI's impact on security and beyond networks from its posture of defender, offender, or victim.

AI's Potential for Cyber-Security in B5G Networks

Massive numbers of linked devices, huge traffic volume, and various technologies (e.g., SDN, NFV) and services will characterize 5G and beyond networks, resulting in a complex and dynamic cyber-threat landscape. With its ability to allow intelligent, adaptive, and autonomous security management, AI adoption is a promising way to deal with this challenging threat scenario.

Identity and Access Management

Authentication and authorization services are critical for 5G and beyond security, as they prevent impersonation attacks and manage access privileges for all parties involved (physical or virtual). Massive machine-type communications (mMTC) and ultra-reliable and low-latency communications (URLLC) use cases, on the other hand, necessitate the support of high device density, energy efficiency, and low-latency capabilities. Meanwhile, in 5G networks, the use of tiny cells will result in more frequent handovers and, as a result, more frequent authentications, resulting in more significant latency. To meet the aforementioned strict requirements, effective, scalable, and quick authentication systems are required. AI is thought to be crucial in reaching this goal. In reality, to determine a given entity's identity and authorizations, developing authentication and authorization systems are increasingly relying on several non-cryptographic factors linked with individuals, resources, and environment (e.g., time and location).

The capacity of AI to automatically combine these different and time-varying attributes to enable continuous authentication and dynamically enforce fine-grained access controls is one of its greatest strengths. Fang et al. introduced ML-based intelligent authentication approaches to achieve continuous and situation-aware authentication in 5G and beyond networks by taking advantage of physical layer attributes (e.g., carrier frequency offset, channel impulse response, and received signal strength indication).

The proposed access control scheme intelligently exploits the time-varying features of the transmitter, such as communication-related, hardware-related attributes, and user behaviors, to refine and update access policies on the fly in a large-scale and dynamic Internet of Things (IoT) environment.

RESTful Application Program Interfaces (APIs) Security

REST APIs (Representational State Transfer APIs) are critical components of the 5G ecosystem since they enable service exposure across multiple networks. As a result, the 3GPP decided that both northbound interfaces (NBIs) and service-based interfaces (SBIs) should be implemented using RESTful APIs. APIs will very indeed become a primary target for attackers due to their importance for 5G and beyond networks. API exploitation/abuse has been listed as a malicious threat to 5G assets in the current ENISA "threat landscape for 5G networks" report, leading to data leakage/alteration/destruction, identity theft, and service unavailability.

As a result, API security is a critical component in protecting 5G and beyond networks. However, identifying and mitigating API vulnerabilities is complex, given the enormous diversity of APIs and the massive amount of API traffic expected in next-generation mobile networks. AI-driven API security is the emerging trend for dealing with the aforementioned difficulties. In reality, AI can identify patterns in massive amounts of multidimensional data, enabling continuous and proactive monitoring and detection of API attacks and automated mitigation.

Note: API security is a broad phrase that refers to procedures and technologies that protect API from malicious attacks or misuse. APIs have become a target for hackers because they are essential for designing web-based interactions.

Network Anomaly/Intrusion Detection and Prediction

Prompt detection and prediction of aberrant behaviors due to malicious or accidental acts is critical to meet the stringent reliability and availability requirements of 5G and beyond networks. Indeed, early detection of possible network problems allows for quick response, preventing severe malicious damage, service degradation, and financial loss. An anomaly refers to a pattern that does not correspond to expected normal behavior.

AI usage has been recognized as a requirement by ETSI Experiential Network Intelligence (ENI) and Industry Specification Group (ISG) to spot anomalous traffic patterns that can lead to service outages or security threats in next-generation networks. AI has demonstrated its capacity to find hidden patterns from many multidimensional data that change over time. An anomaly detection method for Radio Access Networks (RAN) self-healing in 5G networks uses the clustering algorithm DBSCAN to identify abnormal patterns. This method was effective in detecting abnormalities caused by radio attenuation and SDN misconfiguration. The application of shallow and deep learning algorithms to detecting and forecasting network breaches has received a lot of interest. There is a Dynamic Bayesian Network (DBN) model to detect the jamming attack in Orthogonal Frequency Division Multiplexing (OFDM)-based cognitive radio networks.

Root Cause Analysis
The underlying reason for an anomaly alarm must be investigated once it has been triggered. In fact, pinpointing the source of defects and security breaches is essential for empowering self-organizing systems, developing effective mitigation measures, performing network forensics, and even assigning blame. However, given the complexity and heterogeneity of evolving mobile networks and the growing amount of KPIs and data connected to end-users, services, and networks, root cause diagnosis becomes extremely difficult. As a result, a manual evaluation of root causes based on expert knowledge is a complex and time-consuming operation. Because of its ability to handle enormous amounts of data, identify complicated non-linear correlations within the data, and produce faster and more accurate judgments, AI has been identified as an exciting alternative for fostering self-root cause analysis.

AI-driven root cause analysis in the smart manufacturing vertical area can not only assist in finding the source of failure events but also in anticipating future abnormalities, resulting in increased operational efficiency and reduced unplanned downtime.

Moving Target Defense (MTD)

The adversary objective is aided in investigating and attacking the unchanging vulnerability surface by the static nature of network and service configurations once installed. In reality, the persistence of the vulnerability gives the attacker the advantage of time to better understand the attack surface and select the most appropriate assault technique. MTD has developed as a viable proactive security solution for dealing with this issue. No doubt, the National Institute of Standards and Technology (NIST) has designated MTD as an enhanced security criterion for system and communications protection. By dynamically and constantly modifying the attack surface over time, MTD techniques try to increase the attacker's effort and cost. IP address shuffles, virtual machine movement, network path diversity, and software or network resource replication are all examples of MTD implementations. The flexibility and dynamicity afforded by virtualization (Network Function Virtualization [NFV]) and programmability (Software Defined Networking) would encourage the use of MTD methods in 5G and beyond networks, resulting in more resilient networks. The MTD paradigm provides an interesting security technique for numerous vertical application domains, such as IoT and automobiles. A dynamic address/ID shuffling method, for example, can avoid reconnaissance, impersonation, and DoS attacks against in-vehicle networks. Meanwhile, path diversity and topology shuffling can be utilized to improve the eavesdropping and jamming resistance of inter-vehicle wireless communications. It's worth noting that MTD's security benefits come at the cost of reconfiguration and/or service unavailability. As a result, striking the right balance between MTD's security effectiveness and the resulting cost is critical. Game theory, genetic algorithms, and ML have all been mentioned as proper AI methodologies for developing smart MTD mechanisms that intelligently decide what changes to make to the network and service configuration to achieve the security/performance trade-off.

AI's Threats Against B5G Security

Because AI systems will play such a large part in 5G and beyond networks, security issues are an important factor to consider. In fact, the potential threats posed by the use of AI systems can be divided into two categories:

    i. AI as an instrument for developing sophisticated cyber-attacks that leverage AI's capabilities; and

ii. AI as a target for exploiting AI systems' vulnerabilities to undermine their performance and security.

AI as an Instrument

The potential of AI to learn and adapt will pave the way for a new era of autonomous, scalable, stealthy, and speedier AI-powered cyber-attacks. With AI's capabilities combined with the anticipated ultra-high bandwidth and tremendous proliferation of connected devices, AI-driven cyber-attacks will undoubtedly be widely used on 5G and beyond networks. Attackers can employ AI to do a speedy and efficient reconnaissance of the target network, identifying devices deployed, operating systems and services used, ports open, and accounts, particularly those with administrative privileges.

AI can use the information obtained during the reconnaissance phase to learn and prioritize vulnerabilities that could be used to launch a large-scale network attack. For example, an AI-based botnet can detect zero-day vulnerabilities in IoT devices and use them to launch a large-scale DDoS attack against 5G RAN resources by causing a signaling storm.

AI is also likely to be a driving force behind the development of next-generation malware that can work independently. Autonomous malware will be able to observe its surroundings, intelligently identify its target, and use the most effective lateral movement approach to reach it without being detected. DeepLocker is an IBM proof-of-concept autonomous malware that employs a deep neural network model to unleash the harmful payload if the intended victim is found. Various attributes, including geo-location, user activity, and environment data, are used to identify the victim. Another harmful use of AI might be to launch an identity spoofing attack by learning and replicating legitimate companies' behavior.

AI as a Target

AI will be strongly reliant on 5G and beyond networks to provide fully autonomous management capabilities (e.g., self-configuration, self-optimization, self-healing, and self-protection), making AI a tempting target for hackers. In fact, AI systems, particularly ML systems, can be manipulated to develop incorrect models, make inaccurate decisions/predictions, or leak sensitive data. Attacks on ML systems are classified as either causal or exploratory depending on whether they target the training or inference phases. They can be carried out in a white-box, gray-box, or black-box environment, depending on whether the attacker has complete, partial, or no knowledge of the training data, learning algorithm, and hyper-parameters. The adversary may perform indiscriminate attacks to cause the misclassification of any sample or targeted attacks to lead to the misclassification of a specific sample. When attacking an ML system, the adversary may choose to compromise its integrity by avoiding detection while causing the system to behave abnormally; its availability by reducing the system's usability; or its privacy by gaining sensitive information about the training data, the ML system, or its users.

IoT and AI

Today, IoT technology is used to create value through analytics, sensing, and gains from existing connections, among other things. AI/ML can be employed as a critical tool to support IoT applications when it comes to network performance improvement. Data is acquired from numerous areas and industrial sectors by IoT applications. The information gleaned from the data comes from the environments and applications that produced it. AI techniques give the foundation and tools needed to move beyond real-time monitoring and automation use cases in IoT to IoT platforms that apply AI principles to specific IoT use cases to ensure wiser decision-making.

By enabling intelligent automation, predictive analytics, and proactive intervention, AI-enabled IoT applications add a new layer of functionality and access to the next generation of smart homes/buildings, smart vehicles, and smart factories. Deep learning approaches combined with AI, swarm intelligence, and cognitive technologies for maximizing IoT services supplied by IoT applications in smart surroundings and collaboration spaces will generate solutions capable of altering industries and professions. The IoT is a cyber-physical system (CPS).

Note: A CPS can be as simple as a single device, or it can be a system made up of one or more cyber-physical devices, or it can be an SoS made up of several systems made up of many devices. This recurrent pattern is dependent on one's point of view. The decision flow and at least one of the information or action flows must be included in the CPS.

The information flow represents the measurement of the physical state of the physical world in digital form, whereas the action flow has an impact on the physical state of the physical world. This enables partnerships on a local and medium scale all the way up to the city, nation, and global scales.

By merging computation and physical processes, CPSs allow the physical and virtual worlds to blend. In its functioning and interactions with the environment in which it is deployed, a CPS provides for tight integration of computing, communication, and control.

Security in IoT and AI

The IoT promises to connect and integrate ordinary devices such as sensors, actuators, and other physical objects to the Internet, allowing for cutting-edge intelligent services. When it comes to building and developing safe IoT systems, security challenges such as jamming, spoofing, denial of service, eavesdropping, malware in the form of viruses, Trojans, worms, and other threats are a significant cause of concern. They present several possible dangers that might be used to hurt users or even bring an entire system down, such as unlawful access and misuse of personal information, assaults on other systems, and personal safety threats.

Because traditional security mechanisms such as authentication, confidentiality, virus prevention, and so on cannot be directly deployed on IoT devices due to resource constraints, alternative dimensions are assumed in IoT. Traditional compute-intensive security mitigation procedures are impossible to operate on IoT devices due to a lack of resources, battery life, and even network bandwidth. Malicious parties can access and misuse personal information gathered and transferred over IoT devices and networks due to a lack of appropriate security measures, which is a problem that must be addressed immediately. The greater the number of devices linked to the network in a smart home, the greater the number of vulnerabilities a malevolent person could exploit to compromise personal information.

The network is another possible target. Any IoT device can be used to aid attacks against the network to which it is linked and several other connected devices. ML-based AI algorithms, for example, may discover patterns from previous experiences and generate predictions. As a result, AI-based security solutions are likely to respond to new threats more effectively than traditional security approaches. The availability of large data in the security field means that AI approaches may be used to study and recognize patterns of security weaknesses to avoid assaults. An essential element that every IoT system should include is the ability of an IoT-based platform to learn from data, evaluate, recognize, and mitigate security threats.

These methods are also more accurate when it comes to detecting potential malware risks in vast amounts of data. Furthermore, AI is well suited to detecting and mitigating sophisticated attackers such as advanced persistent attacks, which can go undiscovered for an endless amount of time. Insufficient authentication, authorization, insecure network services, lack of transport encryption, insecure cloud and edge interfaces, insecure mobile interfaces, poor security configurability problems, insecure software or firmware, and even poor physical security are all potential security issues in the IoT. It's also worth noting that most IoT devices were built without security in mind because these devices have insufficient processing capabilities to perform security methods.

CHAPTER 12

Road to Future 6G and Security Challenges

It is a fact that the network's focus shifts with each new generation of communications technology. Human-to-human communication via voice and text was the focus of the 2G and 3G periods. The 4G era signaled a fundamental change toward huge data consumption. In contrast, the 5G age is focused on integrating the Internet of Things (IoT) and industrial automation systems.

The digital, physical, and human worlds will fluidly merge in the future 6G to trigger extrasensory experiences. Intelligent information systems will be integrated with powerful processing capabilities to make humans infinitely more efficient and reshape how we live, work, and care for the environment.

1G TO 5G

The cellular wireless Generation (G) typically refers to a change in the appearance of the framework, speed, technology, and frequency. Individually, each generation employs roughly the same standards, capabilities, techniques, and distinctive traits that distinguish it from the previous one (Table 12.1).

First Generation (1G Analog Technology)

The first generation of mobile phones (from 1980 to 1990) worked with data speeds ranging from 1 to 2.8 KBps and used a circuit switch. Analog

Phone Service was the output technology employed. Only the sound will be supported, with a bandwidth of 40 MHz and a frequency range of 800 to 900 MHz. Frequency Division Multiplexing was employed. It made a lot of low-quality calls. The amount of energy consumed was considerable. It suffered from several drawbacks, including a lack of sound connections, a lack of data capacity, a lack of security, and an untrustworthy transfer.

Second Generation (2G Digital Technology)

It is dependent on GSM, or the worldwide mobile communications system, in other words. In 1991, it was marketed in Finland. These were the foremost digital cellular networks, with some similarities to the output networks they replaced in terms of standards and safety. For digital communication, 2G technologies have been superseded by digital technologies that provide text messaging, photo messaging, and MMS services. In 2G technology, all text communications are digitally encoded. This digital encryption allows you to send data that the intended recipient does not understand. The digital modulation techniques used in the 2G network, such as Time Division Multiple Access (TDMA) and Code Division Multiple Access (CDMA), can support both voice and short message services.

Third Generation (3G)

The third generation of mobile transmission networks offers 144 kbps outstanding speeds and more for high-speed data. It complies with advancements in previous wireless technologies, such as "high-speed transmission, high multimedia access, and global roaming." 3G is a technology that connects mobile phones and headphones to the Internet or other IP networks to provide voice and video calls, transfer data, and surf the web. Multimedia applications such as complete video mobility, videoconferencing, and Internet access benefit from 3G. Packet switch technology is used to direct data. A circuit switch decrypts phone calls. It is a very modern communication mechanism that has grown over time.

Fourth Generation (4G)

The IP-based 4G mobile communication architecture was first introduced in the late 2000s. The fundamental goal of 4G developments is to provide high-quality, high-capacity, and low-effort security administrations for phone, data, sound, and Internet services across IP networks. The goal of changing all IP addresses is to create a shared platform for all innovations that have been developed thus far. It has 100 Mbps and 1 Gbps capability. To use the 4G mobile network, multimodal user terminals must carefully select the wireless destination system. Terminal mobility is a key factor in 4G's capacity to supply wireless service anytime, anywhere. Automatic roaming across different wireless networks is suggested by terminal mobility. 4G technology combines various current and future wireless technologies, including "OFDM, MC-CDMA, LAS-CDMA, and Network-LMDS," to provide freedom of movement and seamless roaming from one technology to the next. LTE stands for "long-term evolution," and Wi-MAX is for "wireless interoperability for microwave access." In Japan, the first triumphant fourth-generation field test was organized in 2005.

TABLE 12.1 Summary of Security Evolution from 1G to 4G

Network	Security	Security Challenges
1G	No explicit security and privacy measures.	Eavesdropping, call interception, and no privacy mechanisms.
2G	Authentication, anonymity, and encryption-based protection.	A fake base station, radio link security, one-way authentication, and spamming.
3G	Adopted the 2G security, secure access to the network, introduced Authentication and Kay Agreement (AKA) and two-way authentication	IP traffic security vulnerabilities, encryption keys security, roaming security
4G	Introduced New encryption (EPS-AKA) and trust mechanisms, encryption keys security, non-3G Partnership Project (3GPP) access security, and integrity protection.	Increased IP traffic-induced security, e.g., DoS attacks, data integrity, Base Transceiver Stations (BTS) security, and eavesdropping on long-term keys. Not suitable for security of new services and devices, e.g., massive IoT, Foreseen in 5G

Fifth Generation (5G)

As you know, the fifth generation of cellular networks is known as 5G. Thanks to 5G, which is up to 100 times quicker than 4G, people and businesses will have never-before-seen potential.

Increased bandwidth, ultra-low latency, and faster connectivity expand civilizations, revolutionize industries, and radically improve day-to-day experiences. E-health, networked vehicles and traffic systems, and advanced mobile cloud gaming were formerly considered futuristic.

The advancement of a "World Wide Wireless Web (WWWWW), dynamic ad-hoc wireless networks (DAWN), and actual wireless communication" is the focus of 5G research. The 5G capability gives portable devices AI capabilities. The core network of 5G networks is made up of functions. NFV (Network Function Virtualization), SDN (Software Defined Network), and cloud technologies make networks more dynamic than ever, resulting in a broader range of risks. The comparison of parameters of all generations is provided in Table 12.2.

TABLE 12.2 Comparison of Technology Parameters (1G to 5G)

Technology	1G	2G	3G	4G	5G
Requirements	No official Requirements Analog technology	No official Requirements Digital technology	ITU's IMT 2000 required 144 kbps mobile, 384 kbps pedestrian, and 2 Mbps indoors	ITU's IMT Advanced requirements include the ability to operate in up to 40 MHZ radio channels and with very high spectral Efficiency.	At least 1 GB/s or more data rates to support ultra-high definition video and virtual reality, applications, 10 GB/s data rates to support mobile cloud service
Data Bandwidth	1.9 kbps	14.4 kbps– 384 kbps	2 Mbps	2 Mbps– 1 Gbps	1 Gbps & Higher (as demand)
Core network	PSTN	PSTN Packet Network	Packet Network	All IP Network	Flatter IP Network & 5G Network Interfacing (5G-NI)
Services	Analog voice	Digital voice Higher capacity, packetized data	Integrated high-quality audio, video, and data	Dynamic information access, wearable devices, HD streaming; global roaming	Dynamic information access, wearable devices, HD streaming; any demand of user; upcoming all technologies; global roaming smoothly;
Standards	NMT, AMPS, Hicap, CDPD, TACS, ETACS	GSM,SPRS, EDGE ETC.	WCDMA, CDMA 2000.	All-access convergence, including-OFMDA, MC-CDMA Network-LMPS	CDMA & BDMA
Starts from	1970–84	1990	2001	2010	2015
Switching	Circuit	Circuit Packet	Circuit Packet	Packet	All Packet
Frequency	800–900 Reference Mhz	850–1900 MHZ	1.6–2.5GHZ	2–8GHZ	

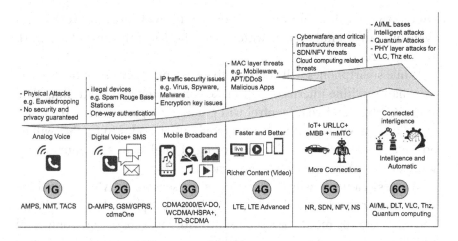

FIGURE 12.1   Evolution of communication network security landscape.

6G ERA BENEFITS

Every increase in network connectivity that 5G brings to the end-user will be refined even more with 6G. 6G will push smart cities, farms, factories, and robotics to the next level. It will be made possible by 5G-Advanced, the next generation of 5G standard upgrades. It has increased efficiency, expanded capabilities, and a better user experience.

Looking back, it's evident that each generation improves on the preceding generation's use cases while also introducing new ones. This will be the situation in the future. 6G will be built on top of 5G in many technological and use-case aspects, enabling widespread adoption through cost-cutting and optimization. At the same time, 6G will open up new possibilities.

The enormous scale deployment of sensors, artificial intelligence, and machine learning (AI/ML) with digital twin models and real-time synchronous updates will connect the physical world to our human world. These digital twin models are critical because they allow us to examine what's going on in the real world, simulate probable outcomes, predict needs, and then return to the physical world to take productive actions.

With 5G, digital twin models are already in use. We may anticipate these technologies operating on a much larger scale with 6G.

Digital twins will be found in factories, city-wide area networks, and even human digital twins, which will significantly impact network architecture.

While smartphones will continue to be essential devices in the 6G age, new man-machine interfaces will make consuming and controlling data easier. Gesture and voice control will gradually replace touchscreen typing. Clothing will have devices integrated into it, and skin patches will be possible. Wearables allow the monitoring of vital factors 24 hours a day, seven days a week, which will benefit healthcare.

Wireless cameras will become universal sensors as AI and machine vision mature and have the ability to detect people and objects. Radio and other sensing modalities, such as acoustics will collect environmental data. It's possible that digital cash and keys may become the norm. We may even begin to use brain sensors to control machinery.

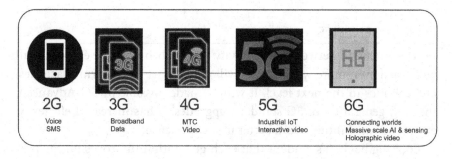

FIGURE 12.2   Mobile Communication from 2G to 6G.

In addition, 6G will boost sustainability in a number of ways. It would be able to support data collecting and closed-loop control of various appliances by enabling faster and cheaper cost-per-bit communication. Advanced tools can be used to analyze the data to enhance energy efficiency in enterprises. Through the advent of multimodal mixed reality telepresence and remote collaboration, enhanced multi-sensory telepresence generated with very high data rates would minimize the need for travel.

When demand is low, 6G will turn off components and scale down capacity, making it substantially more energy-efficient. Along with other criteria like capacity, peak data rate, latency, and reliability, energy efficiency will be an important design criterion in 6G.

THE 6G NETWORK

The 6G cellular technology is the successor to the 5G cellular technology (sixth-generation wireless). 6G networks will be able to operate at greater frequencies than 5G networks, resulting in considerable capacity and latency improvements. One of the 6G internet's goals is to achieve communication with a one-microsecond latency. One-millisecond throughput (or 1/1000th the latency) is 1,000 times faster.

The market for 6G technology is predicted to enable significant advancements in imaging, presence technologies, and location awareness. The computational infrastructure of 6G will autonomously select the ideal place for computing, including decisions regarding data storage, processing, and sharing, using AI.

6G will necessitate a shift in the way communication networks are built. Multiple significant objectives must be reconciled: serve exponentially increasing traffic and an ever-increasing number of devices and markets while meeting the highest possible standards in terms of performance, energy efficiency, and robust security, allowing for sustainable growth in a secure manner.

5G-Advanced is a critical stepping stone toward some of the features we want to enable in 6G on a bigger scale. Over the next half-decade, it will further develop 5G to its full potential. How networks are architected, constructed, and deployed in the 5G-Advanced era to cope with traffic growth will require a new level of intelligence. One that can be controlled across a disaggregated network and powered by AI and Closed-Loop Automation. The transition to 5G-Advanced will also necessitate the best support for essential network applications, whether through communication service providers (CSPs) or industry-grade private wireless networks.

SIX TECHNOLOGY AREAS WILL CHARACTERIZE 6G

According to Nokia Bell labs, six technologies are going to empower and characterize 6G.

AI and ML

AI/ML techniques, particularly deep learning, have grown significantly over the previous decade. They have already been applied in various fields involving picture categorization and computer vision, ranging from social networks to security. With the techniques in 5G-Advanced, AI/ML will be integrated into many elements of the network at many layers and in many functions, allowing these technologies to reach their full potential. From beam-forming optimization in the radio layer to self-optimizing networks for cell site scheduling, AI/ML is being used to provide higher performance with less complexity.

Spectrum Bands

In order to provide radio communication, spectrum is essential. Every new mobile generation necessitates a new pioneer spectrum that allows a new technology's full potential to be realized. It will also be essential to reform the existing mobile communication spectrum from legacy technology to the next generation. Mid-bands 7–20 GHz for urban outdoor cells enabling higher capacity through extreme multiple-input, multiple-output (MIMO), low bands 460–694 MHz for extreme coverage, and sub-Terahertz(THz) for peak data rates that exceeds100 Gbps are likely to be the new 6G pioneer spectrum blocks.

While 5G-Advanced will expand 5G beyond data communication and significantly improve positioning accuracy to centimeter-level, especially indoors and underground where satellite signals are unavailable, 6G will take localization to the next level by utilizing a wide spectrum and new spectral ranges up to THz.

A Network That Can Sense

The ability of 6G to perceive the environment, people, and objects is its most prominent feature. The network becomes a source of situational data, collecting signals bouncing off objects and detecting their type and shape, relative location, velocity, and possibly even material qualities. In combination with other sensing modalities, such a form of sense can help build a "mirror" or digital counterpart of the physical world, extending our sensations to every point the network touches. By combining this data with AI/ML, the network will gain fresh insights from the physical world, making it more cognitive.

Extreme Connectivity

The Ultra-Reliable Low-Latency Communication (URLLC) service, which debuted with 5G, will be enhanced and improved in 6G to meet the most stringent connectivity standards, including sub-millisecond latency. Simultaneous transmission, numerous wireless hops, device-to-device connections, and AI/ML could improve network stability. Real-time video communications, holographic experiences, and even digital twin models updated in real-time through the deployment of video sensors will benefit from improved mobile broadband combined with decreased latency and increased reliability.

We can expect use cases with networks that have unique requirements in sub-networks in the 6G era and networks of networks with networks as an endpoint. Machine area networks, such as a car or body area network, might have hundreds of sensors spread out across a small region of fewer than 100 meters. These sensors must communicate within 100 microseconds with extremely high reliability for the machine system to operate. The creation of wireless networks in automobiles and robots will usher in a new era for device designers since they will no longer be required to build long, unwieldy cable systems.

New Network Architectures

5G is the first solution meant to replace wired connectivity in the enterprise/industrial sector. The industry will require ever more complex designs that can allow increased flexibility and specialization as demand and strain on the network increase.

The network will be deployed in heterogeneous cloud settings incorporating a mix of private, public, and hybrid clouds. 5G will introduce services-based architecture in the core and cloud-native deployments that will be extended to parts of the Radio Access Network (RAN). Furthermore, as the core gets more dispersed and the RAN's higher layers become more centralized, there will be chances to save money by combining functions. New network and service orchestration solutions that take advantage of AI/ML developments will result in unparalleled network automation and cost savings.

Security and Trust

Cyber-attacks are growing more common on all types of networks. The ever-changing nature of the threats necessitates the deployment of robust security procedures. 6G networks will be built to withstand threats such as jamming. When creating new mixed-reality worlds that combine digital representations of actual and virtual objects, privacy considerations will need to be considered.

FIGURE 12.3   Key technologies for 6G.

RESEARCH ON 6G IS ON THE WAY

While we are currently experiencing the active deployment of 5G commercialization, now is the ideal moment to begin planning for 6G, which will become a part of our lives in 2030 and beyond. Samsung Development established the Advanced Communications Research Center in 2019 to boost 6G research.

Several research on 6G technology have identified the following megatrends in the direction of 6G: As shown in Figure 12.4, network openness, network AI (growth of AI usage), and massively connected machines/robots are all important.

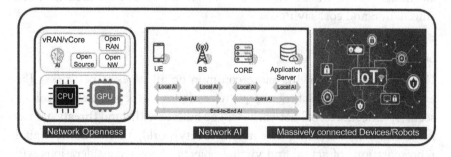

FIGURE 12.4   6G megatrends.

Trustworthiness is believed to be a fundamental criterion of 6G, in addition to traditional performance requirements such as data rate and latency, which have been considered in previous generations. We must build 6G as the most secure mobile communication technology ever.

Network openness, SW (software)-based mobile communication systems, virtualization/containerization, network AI, and quantum computing are all technological advancements that directly affect security and privacy. To make 6G mobile communication secure, new hazards associated with each technology trend should be investigated.

6G SECURITY THREATS FROM TECHNOLOGY TRENDS

Security threats (Table 12.3) are predicted to emerge due to significant technological advances toward 6G.

Network Openness Threats

As low-performance devices such as drones, household appliances, and smart sensors grow exponentially, the system may become more susceptible in an environment where devices with inadequate security reliability access mobile communication systems.

Open Source SW Threats

Using open-source software increases the attack surface. Year after year, the number of vulnerabilities in open-source software grows, from 500 in 2008 to over 6,000 in 2019. An attacker can easily select an attack and analyze the target's operation because the source code is exposed. In particular, the open-source software that isn't regularly maintained may be much more vulnerable because fixes are frequently delivered insufficiently or incorrectly.

Threats from Network AI

Using AI's training and interference properties, an attacker can purposefully confuse an AI model by entering fraudulent data, causing the network system to become unstable, dysfunctional, or unavailable, known as Adversarial Machine Learning (AML).

Privacy Threats

To improve quality-of-service (QoS) and quality of experience (QoE), mobile network operators (MNOs) and external service providers are

TABLE 12.3 Overview of Main Security and Privacy Issues in Key 6G Technologies

Key Technology	Security and Privacy Issues	Key Technology Contribution
AI	Access Control malicious behavior Authentication Communication Encryption	-Fine-grained control processes Detect network anomalies and provide early warning -An unsupervised learning method that could be used in the authentication process to enhance the security of the physical layers -An antenna design based on machine learning that could be used in PHY layer -Communication to prevent Information leaks -Machine learning and quantum -Encryption schemes
Molecular Communication	Malicious Behavior Encryption Authentication	-An adversary disrupting Molecular communication or its Processes -A coding scheme that could enhance the security of data transmission -Provides direction for developing new authentication mechanisms
Quantum communication	Encryption Communication	-Mechanisms for protecting quantum encryption keys -Different modes of quantum communication
Blockchain	Authentication Access control Communication	-A new conceptual architecture for mobile service authorization -A method that improves access protocols -Using hashing power to validate transactions
THZ	Authentication Malicious Behavior	-Electromagnetic signatures as a method of authentication -Even when a signal is transmitted via a narrow beam, -It can still be intercepted by an Eavesdropper
VLC	Communication Malicious Behavior	-A secure protocol that can be used in the communication process -Cooperating eavesdroppers can reduce security

required to analyze user data in a variety of ways using data analytics and AI technology. On the other hand, personalization technologies in 6G open the door to fine-grained privacy infringement. As a result, if a trustworthy and secure environment is not in place, the use of user information may leak and pose significant dangers to user privacy.

Virtualization/Containerization Threats

Virtualization and containerization are technologies that allow a single physical hardware (HW) resource to be partitioned into many logical resources as independent virtual machines (VMs) or containers, with the operating system (OS)-level isolation of environments for running software services. Operators benefit greatly from virtualization/containerization in terms of capital expenditure (CAPEX) and operating expenditure (OPEX). Vulnerabilities can readily propagate across all VMs or containers unless the hypervisor maintains VMs or the operating system is securely supported.

Threats Associated with Quantum Computing

Quantum computing (QC) uses quantum mechanical phenomena like superposition and entanglement to solve computationally difficult tasks like prime factorization of a big number and the problem of discrete logarithms. Some encryption methods will no longer be secure if QC becomes more commonly used in the 6G era.

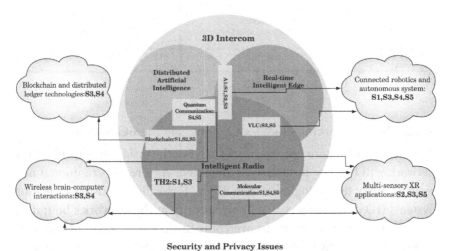

Security and Privacy Issues

S1: Authentication, S2:Access Control, S3: Malicious behaviors, S4: Encryption, S5: Communication

FIGURE 12.5   Security and privacy issues in the 6G network.

TOWARD 6G SECURITY: SOLUTIONS FOR TRUSTWORTHINESS

We expect trustworthiness to become one of the essential needs in 6G, given the increased danger of security threats. Table 12.4 summarizes potential options for trustworthiness provisioning, and we go over some of them in more detail below.

TABLE 12.4 6G Security: Trustworthiness Solutions for Various Security Threats

Technology Trends	Security Threats	Solutions
Network openness	Unauthorized access or access bypass Machines with low security Unauthorized access in the open interface SW collision	PKI and zero trust architecture (ZTA) Security governance system Identification and authentication management Conflict migration
Open Source SW-based Mobile Communication System	Undetected vulnerability Unpatched open source	Automated vulnerability management system Automatic patch system, secure OTA
Virtualization and containerization	Compromized hypervison Privilege escalation Inter-VM attack Vulnerability spread with live migration Unauthorized access, packet sniffing on socket, etc.	Trusted execution environment (TEE) Secure Virtualization layer Virtual machine introspection (VMI) Secure live migration Secure configuration at OS Level
Utilization of User Information	Unauthorized access Unauthorized usage Exposure of user information UE ID catching attacks	Secure authentication and authorization Secure procedures for usage of user information Trusted execution environment (TEE) Secure ID assignment
Network AI	Malfunction of AI Model inversion attack Data poisoning attack Model evasion attack Model extraction attack	Trustworthy AI inference systems AI-training with privacy-enhancing technologies Reliable AI-data collection Statistical analysis of AI operation Threshold-based anomaly detection
Introduction of quantum computing	Unsecure asymmetric key algorithm	Adopting of post-quantum cryptography (PQC)

System Security Architecture for Openness in the 6G System

As 6G becomes a more open network than 5G, the inside and outside of the network will become progressively muddled. As a result, traditional network security technologies such as IPsec, Firewalls, Intrusion Detection Systems (IDS), and others that apply security at the network's boundary point would be insufficient. To overcome this constraint, the 6G security architecture should support the mobile communication network's core security premise of zero trust (ZT). It is a security paradigm that focuses on system resource protection. ZT considers that an attacker could be present within the network and that network infrastructure is accessible or untrustworthy from the outside. As a result, it is vital to continually assess the trustworthiness of internal assets in the face of threats and to take precautionary actions to limit the risks.

As shown in Figure 12.6, zero trust architecture (ZTA) is a security architecture that incorporates the ZT concept and comprises relationships between network entities (NEs), protocol procedures, and access controls.

As a result, in a 6G security architecture, ZTA should be the primary security principle.

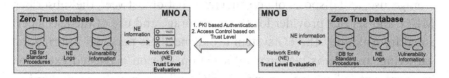

FIGURE 12.6   Illustration of zero trust concept.

Automated Management System for Open Source Security

The most significant thing to do to address the open-source security issue is to employ an automated management system during the development of the 6G system, which handles vulnerabilities caused by open-source use, update, and disposal. An automated management system to detect vulnerabilities and deploy a patch is required to check the threats swiftly. In addition, a secure over-the-air (OTA) mechanism must be implemented to ensure that the patched software is applied as soon as possible. Furthermore, a security governance framework must be adopted through a specialized organization to monitor 1) long-term open source vulnerabilities, 2) changes in developer perception, and 3) the adoption of security solutions.

Solution for Virtualization/Containerization Security

Due to different virtualization/containerization security difficulties, systems must only be run on a system with a safe virtualization layer that includes a security solution that identifies hidden harmful software such as rootkits. Furthermore, the hypervisor must enable total isolation of computing, storage, and network of various network services using secure protocols such as transport layer security (TLS), secure shell (SSH), a virtual private network (VPN), and others. Virtual machine introspection (VMI) examines and detects security threats. Threats are based on the vCPU register information, vMem data, file IO, and communication packets of each VM, and the hypervisor must be equipped with IDS/IPS/firewall to study the internal state of the VM and identify an intrusion. In the case of containerization, the operating system should appropriately configure the privileges of various containers, prohibiting the mounting of main system directories and direct access to the host device file container.

Privacy-Preserving Solution

Mutually agreeable protocols should be set throughout the collection, storage, use, and disposal of personal information between the subscriber, MNO, and service provider for the safe storage and use of the user's personal information. The 6G system avoids needless personal information gathering, maintains personal information safe on a trusted execution environment (TEE) and dependable SW, and minimizes or anonymizes exposed information while using personal information through these procedures. MNO shall verify the authenticity and authorization of personal information before delivering it. Furthermore, while accessing and analyzing user data, it can be anonymized to safeguard privacy or encrypted using homomorphic encryption (HE) to make the data available in an encrypted manner. To secure both the user location privacy and the usage pattern privacy, AI-based solutions such as learning-based privacy-aware offloading schemes can be used.

Note: Homomorphic encryption is a cryptographic technique that allows mathematical operations on data to be performed on cipher text rather than the original data.

AI Security Technologies

Transparency assessing how securely AI systems work against AML must be ensured to protect subscribers and mobile communication systems from AML. To begin, AI models must be generated in a secure environment. It's also necessary to adopt a technique like a digital signature to ensure that AI models in user equipment (UE), RAN, and core haven't been maliciously modified or altered by a malicious attack. A system must perform self-healing or recovery methods if a hostile AI model is found. To ensure that AI is as reliable as possible, the system should limit data collection for AI training to only trustworthy network nodes.

Note: AML is the process of extracting information about the behavior and characteristics of an ML system and/or learning how to manipulate the inputs into an ML system to obtain a preferred outcome.

Adoption of Post-Quantum Cryptography

Existing asymmetric key encryption schemes, which will become insecure with the onset of quantum computing, should be phased out of the 6G system. Many academics are working on Post-Quantum Cryptography (PQC) solutions, such as lattice-based cryptography, code-based cryptography, multivariate polynomial cryptography, and hash-based signatures.

Between 2022 and 2024, the US National Institute of Standards and Technology (NIST) oversees PQC research and is expected to identify appropriate PQC algorithms. The length of the key currently being considered for PQC is projected to be several times longer than the 1000 bits of the commonly used Rivest–Shamir–Adleman (RSA) algorithm. PQCs are projected to have a higher computational overhead than the present RSA technique. As a result, it's critical to align PQC with the 6G network's HW/SW performance and service needs.

Note: PQC stands for "post-quantum cryptography." It refers to a collection of classical cryptographic methods that are thought to be "quantum-safe," meaning that they should remain secure even in the presence of quantum computers.

PROMISING TECHNOLOGIES FOR EXPLORATION

Four candidate technologies shine out in terms of business opportunity and viability when considering 6G's prospects and promise.

Joint Communication and Sensing

Increased data and more ambient sensing and awareness are required for the 6G experience, and joint communications and sensing investigate how to combine them. Autonomous vehicles, for example, have extremely complex sensing systems that combine data from a variety of cameras, lidar, and radar sensors using ML algorithms. These vehicles' advanced communications systems rely on cellular networks for streaming infotainment, environment and performance data, and vehicle-to-everything connections.

Those working on sensing are looking to communications technologies like orthogonal frequency division multiplexing (OFDM) waveforms or MIMO phased arrays to help them improve their results, while those working on communications see the vast swaths of the radar-allocated spectrum as an opportunity for more data bandwidth. Legislative and technical constraints will determine the amount to which these two historically distinct activities combine, but the result could define 6G.

Note: Vehicle to Everything (V2X) is a vehicular communication system that allows information to be transferred from a vehicle to moving traffic parts.

Sub-THz

Researchers are exploring unused spectrum in the sub-THz frequency regions as a result of the constant demand for increased data bandwidth.

Many times, the spectrum now used for cellular communications is available in frequency bands between 90 and 300 GHz. The 3GPP has already identified 21.2 GHz above 100 GHz as a suitable 6G frequency.

Path loss at higher frequencies, one of the most significant barriers to migrating to sub-THz bands, can be minimized by matching the attenuation features of a frequency band with appropriate applications like high-attenuation bands for high-security applications and limiting how far the signal travels. Additionally, one technique to avoid path loss is to use the inverse relationship between frequency and antenna size: As frequency rises, antenna shape and spacing decrease, allowing more elements and hence more gain to be packed into the same space. While moving to sub-THz bands may seem premature given the slow pace of 5G mmWave deployments so far, major industry and university researchers are looking into it as a way to boost network capacity greatly.

Evolution of MIMO

MIMO continues to improve on current multiantenna approaches, with potential across a wide range of use cases and frequency ranges. Multiuser MIMO considerably improves spectral efficiency for the most widely used sub-8 GHz bands, while beam-forming is critical for overcoming sub-THz path loss issues. Distributed MIMO, which divides huge antenna arrays into numerous smaller, geographically separated radio heads, is particularly appealing for frequencies below 8 GHz, where antenna size becomes prohibitively vast. The expansion of MIMO to encompass greater system antenna counts for more users and more accurately guided beam steering promises to boost cell capacity and improve location services.

Note: Path loss (PL) or path attenuation describes how the power density of an electromagnetic wave decreases as it travels across space.

AI and ML

AI and ML are the fourth technology that will undoubtedly play a big role. Traditional signal-processing approaches become increasingly difficult to optimize the communications system as complexity rises and we try to squeeze every last bit of bandwidth out of the available spectrum. One method to deal with this complication is to use ML. Automatic spectrum allocation, beam management, and RF nonideality cancelation could benefit AI/ML-driven design or adaptation to optimize connection performance dynamically. By deploying AI/ML at the application layer, QoS can be improved, which considers application-specific requirements and the environment for aspects like latency and energy efficiency. The availability of large, free datasets for AI/ML wireless communication research and training will be critical to the advancement of 6G.

EXPLOITING THE AIRWAVES FOR 6G

Every iteration of mobility has boosted capacity by increasing carrier bandwidth, boosting the number of airwaves available for data transmission. Carrier size increased from 5 to 20 MHz as 3G to 4G was upgraded, while carrier bandwidth increased from 20 to 100 MHz as 4G to 5G was upgraded. When 6G becomes available, we predict spectral bandwidths to grow once again, hitting 400 MHz, considerably enhancing a single cell's baseline capacity.

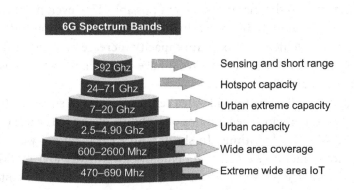

FIGURE 12.7   The spectrum wedding cake.

We'll have to carve out vast blocks of frequencies from previously untapped spectrums to accomplish these large carrier sizes. As 5G has gained traction, governments worldwide have realized the importance of improved mobile networks in driving national economic growth, providing them with ample motivation to clear more spectrum for mobile use. New frequency bands between 7 and 20 GHz are expected to open up for 6G use in ten years, providing the necessary bandwidth to develop these new high-capacity carriers.

Mobile networks will use this so-called mid-band 6G spectrum as their workhorse frequency. Lower band bandwidth allocations are insufficient to build the 400-MHz carriers required for 6G. Meanwhile, the ultra-high frequency bands are prohibitively expensive to use for wide-area coverage due to the difficult signal propagation circumstances. Mid-band spectrum is ideal for delivering high capacity while maintaining competitive coverage.

These frequencies, however, will not be the primary source of additional spectrum for 6G networks. We see governments continuing to identify more bands for mobile use, further accelerating the mobile broadband revolution. In the 6G future, it's also conceivable that a spectrum will be provided for new types of use cases. Regulators are looking at the 470–694 MHz spectrum, for example, as a way to provide broad coverage in rural and remote areas. This band's signals propagate significantly further due to the low frequencies, extending the network's reach. Beyond 90 GHz, sub-THz bands could provide exceptionally high peak data rates for the most bandwidth-intensive applications and connect exceedingly dense sensor networks. Using extreme massive MIMO to reuse airwaves to the maximum extent possible. However, having a new spectrum isn't enough. By increasing carrier bandwidth from 100 to 400 MHz, we may achieve a maximum capacity increase of 4×, well short of the 20× demand of the 6G era. We'll have to find new ways to use and reuse that spectrum.

The spectral efficiency of wide-area cells has improved significantly during the several previous generations of mobility, owing to the use of more advanced MIMO algorithms. In short, with each iteration, we've introduced more antenna elements: 4G employs 2×2MIMO and 4×4MIMO, but 5G uses large MIMO with up to 64 transceivers and 200 antennae components. 6G could accommodate up to 1024 antenna elements in the new mid-bands.

We can handle denser antenna arrays by shifting into higher frequency bands. Because the 6G mid-bands have double the frequency of today's 3.5 GHz 5G bands, we can fit four times as many antennas onto a similar-sized antenna array. By broadcasting many more streams of data simultaneously, these enormous antenna arrays will assist achieve a significant boost in capacity.

To make these high-performance arrays a reality, significant technological advancements are necessary. New scalable, low-power radio-frequency and digital front ends, as well as more complex beam-forming algorithms and high-capacity front haul and baseband processing, will be required.

THE ROADMAP TO 6G IS AI-EMPOWERED WIRELESS NETWORKS

While 5G is still in its beginning, it is essential for businesses and academia to consider what 6G will entail in order to ensure the long-term viability and competitiveness of wireless communication networks. There are already initiatives in detailing the 6G roadmap, developing trends and requirements, and numerous enabling techniques and architectures, such as THz band communications. Unlike earlier generations, 6G will be transformational, revolutionizing the wireless development from "connected objects" to "connected intelligence," with the following more demanding requirements.

- Extremely high data speeds of up to 1 Tbps;

- Extremely great energy efficiency, allowing for battery-free IoT devices;

- Global connectivity that can be trusted;

- Massive control with minimal latency

- A wide range of frequency bands

- Integrated terrestrial wireless and satellite systems provide ubiquitous, always-on broadband global network coverage;

- AI networking hierarchy and connected intelligence with ML capabilities. In addition to the eMBB, uRLLC, and mMTC services offered by 5G, 6G will require the support of three more service categories, as stated below.

Computation Oriented Communications (COC): New smart devices require distributed and in-network computation to allow the main functions of AI-powered 6G, such as federated learning and edge intelligence. Instead of focusing on traditional QoS provisioning, COC will determine an operating point in the rate-latency-reliability space based on the availability of various communications resources in order to attain a given level of computational accuracy.

Note: Federated learning is an ML technique that involves training an algorithm through several decentralized edge devices or servers that carry local data samples without sharing them. Edge artificial intelligence (edge AI) is a paradigm for crafting AI workflows that span centralized data centers (the cloud) and devices outside the cloud that are closer to humans and physical things (the edge).

Contextually Agile eMBB Communications (CAeC): Communication network context, such as link congestion and network topology; physical environment context, such as surrounding location and mobility; and social network contexts, such as social neighborhood and sentiments, are expected to make 6G eMBB services more agile and adaptive to the network context.

Event Defined uRLLC (EDuRLLC): Unlike the 5G uRLLC application scenario like virtual reality and industrial automation, where redundant resources are in place to mitigate many uncertainties. 6G will need to support uRLLC in extreme or emergency events with spatially and temporally changing device densities, traffic patterns, spectrum, and infrastructure availability.

Note: A part of the 5G network design, Ultra-Reliable Low Latency Communications (URLLC) ensures more efficient data transfer scheduling, accomplishing shorter transmissions across a bigger subcarrier, and even scheduling overlapping transmissions.

AI-ENABLED TECHNOLOGIES FOR 6G

6G will differ from previous generations with a high degree of heterogeneity in several facets, such as network infrastructures, RF devices, radio access technologies, computing and storage resources, application types, etc., due to the unprecedented transformation of wireless networks.

Furthermore, a diverse set of new applications will necessitate the intelligent utilization of communications, computation, control, and storage resources taken from the network edge to the core and then across numerous radio technologies and network platforms. Finally, the volume and variety of data created through wireless networks are rapidly increasing. This gives up a lot of possibilities for data-driven network planning and operation in dynamic network environments to achieve real-time additivity. In this section, we argue that AI is an essential tool for intelligent learning, reasoning, and decision-making in 6G wireless networks.

Big Data Analytics for 6G

Big data analytics is the first natural use of AI. Descriptive, diagnostic, predictive, and prescriptive analytics are the four forms of analytics that can be used with 6G systems.

Descriptive analytics mines historical data for information about network performance, traffic profiles, channel conditions, user viewpoints, and so on. It considerably improves network operators' and service providers' situational awareness.

Diagnostic analytics enable autonomous detection of network failures and service impairments and identification of the fundamental causes of network anomalies, resulting in improved 6G wireless system reliability and security.

Predictive analytics uses data to forecast future events such as traffic patterns, user locations, behavior and preference, content popularity, and resource availability.

Prescriptive analytics uses predictions to advise resource allocation, network slicing and virtualization, cache placement, edge computing, autonomous driving, and other decision possibilities. For example, proactive caching has recently emerged to dramatically relieve peak traffic loads from the wireless core network by forecasting, anticipating, and inferring future user requests through big data analytics.

AI-Enabled Closed-Loop Optimization

Traditional wireless network optimization approaches may not be helpful in 6G systems for the following reasons. First, because of the network's scale, density, and heterogeneity, 6G wireless networks will be tremendously dynamic and complicated. It is challenging, if not impossible, to model such systems. As a result, existing optimization methods that primarily rely on mathematically convenient models will be insufficient. As a result, automated and closed-loop optimization is the second significant use of AI in 6G wireless systems. Traditionally, problems in wireless networks have been tackled by combining sets of rules generated from system analysis with domain knowledge and experience. Traditional network optimization issues, for example, presume that the objective functions are given in suitable algebraic forms, which allows an optimizer to evaluate a solution with a straightforward calculation. However, in the complex 6G network environment, defining the relationship between a decision and its impact on the physical system is costly and may not be possible analytically. Recent advancements in AI technologies, such as reinforcement learning and deep reinforcement learning (DRL), can create a feedback loop between the decision-maker and the physical system. It will allow the decision-maker to iteratively refine his or her action based on the system's feedback until it reaches optimality. For example, DRL was recently used to handle a number of developing communication and networking difficulties, such as adaptive modulation, wireless caching, data offloading, and so on.

Intelligent Wireless Communication

AI technologies will play a crucial role in the end-to-end optimization of the physical layer signal processing chain from the transmitter to the receiver. Hardware issues, including amplifier distortion, quadrature imbalance, local oscillator, clock harmonic leakage, and channel issues like fading and interference, plague the end-to-end communication system. Meanwhile, the number of variables and parameters that must be managed will continue to grow. End-to-end optimization has never been feasible in today's wireless systems due to this level of complexity. Existing techniques, on the other hand, break the entire chain into numerous isolated blocks, each with a simplified model which does not accurately or holistically capture the characteristics of real-world systems. AI technologies enable us to learn the optimum method to communicate using a variety of hardware and channel impacts. In 6G, we foresee an "intelligent PHY layer" paradigm, in which the end-to-end system can learn and optimize itself by combining sophisticated sensing and data collecting, AI technologies, and domain-specific signal processing methodologies.

Bibliography

1. https://www.3gpp.org/
2. https://5g.systemsapproach.org/arch.html
3. https://www.etsi.org/deliver/etsi_ts/133500_133599/133501/15.02.00_60/ts_133501v150200p.pdf
4. https://www.ericsson.com/en/reports-and-papers/
5. https://www.5gamericas.org/security-for-5g/
6. https://www.5gamericas.org/security-considerations-for-the-5g-era/
7. https://www.5gamericas.org/5g-and-the-cloud/
8. https://cltc.berkeley.edu/wp-content/uploads/2020/09/Security_Implications_5G.pdf
9. https://www.enisa.europa.eu/publications/enisa-threat-landscape-2021
10. https://www.enisa.europa.eu/publications/security-in-5g-specifications
11. https://www.researchgate.net/publication/345761252_5G_In_the_internet_of_things_era_An_overview_on_security_and_privacy_challenges
12. https://www.researchgate.net/publication/318223878_5G_Security_Analysis_of_Threats_and_Solutions
13. https://arxiv.org/ftp/arxiv/papers/1506/1506.02876.pdf
14. https://www.sdxcentral.com/5g/
15. https://www.viavisolutions.com/en-us/5g-network-deployment
16. https://www.kaspersky.com/resource-center/definitions/ai-cybersecurity
17. https://webthesis.biblio.polito.it/12557/1/tesi.pdf
18. https://www.mdpi.com/1424-8220/20/3/828/htm
19. https://www.cablelabs.com/insights/a-comparative-introduction-to-4g-and-5g-authentication
20. https://www.tech-invite.com/3m23/toc/tinv-3gpp-23-501_zy.html#e-6-2-5-0
21. https://www.techrepublic.com/
22. https://www.techplayon.com/5g-authentication-and-key-management-aka-procedure/
23. https://www.cablelabs.com/insights/a-comparative-introduction-to-4g-and-5g-authentication
24. https://research.samsung.com/blog/Towards-6G-Security-Technology-Trends-Threats-and-Solutions
25. www.nokia.com/blog
26. www.techtarget.com

27. www.bell-labs.com
28. https://www.ericsson.com/en/blog/2020/6/security-standards-role-in-5g
29. https://www.orfonline.org/wp-content/uploads/2021/12/ORF_Monograph-ANU-Quad-5GResilience.pdf
30. https://education.dellemc.com/content/dam/dell-emc/documents/en-us/2020KS_Gloukhovtsev_How_5G_Transforms_Cloud_Computing.pdf
31. http://5gensure.eu/sites/default/files/5G-ENSURE_D2.7_Security ArchitectureFinal.pdf
32. https://www.ngmn.org/wp-content/uploads/210804-NGMN-Security-Considerations-for-5G-Network-Operation-V1.0.pdf
33. http://jultika.oulu.fi/files/nbnfi-fe2021081643336.pdf
34. https://www.nokia.com/networks/5g/mobile/5g-resources/
35. https://www.gartner.com/smarterwithgartner/the-cios-guide-to-distributed-cloud
36. "A Survey on Security and Privacy of 5G Technologies: Potential Solutions, Recent Advancements and Future Directions" Rabia Khan, Student Member, IEEE, Pardeep Kumar Member, IEEE, Dushantha Nalin K. Jayakody*, Senior Member, IEEE and Madhusanka Liyanage, Member, IEEE
37. "Security for 5G Mobile Wireless Networks" DONGFENG FANG1, YI QIAN 1, (Senior Member, IEEE), AND ROSE QINGYANG HU2, (Senior Member, IEEE), Department of Electrical and Computer Engineering, University of NebraskaLincoln, Omaha, NE 68182, USA, Department of Electrical and Computer Engineering, Utah State University, Logan, UT 84321, USA
38. "The Roadmap to 6G Security and Privacy" PAWANI PORAMBAGE (Member, IEEE), GÜRKAN GÜR (Senior Member, IEEE), DIANA PAMELA MOYA OSORIO (Member, IEEE), MADHUSANKA LIYANAGE (Senior Member, IEEE), ANDREI GURTOV (Senior Member, IEEE), AND MIKA YLIANTTILA (Senior Member, IEEE)
39. "Security Risk Prevention and Control Deployment for 5G Private Industrial Networks" Wenfa Yan1, Qin Shu, Peng Gao
40. "5G Security Challenges and Opportunities: A System Approach", Ashutosh Dutta *Applied Physics Lab Johns Hopkins University*, Baltimore, MD, USA, Eman Hammad *Department of Computer Science Texas A&M University - Commerce/RELLIS* College Station, TX, USA
41. "Security and privacy in 6G networks: New areas and new challenges" Minghao Wang, Tianqing Zhu, Tao Zhang, Jun Zhang, Shui Yu, Wanlei Zhou School of Computer Science, University of Technology Sydney, Ultimo, 2007, Australia School of Software and Electrical Engineering, Swinburne University of Technology, Australia
42. "5G Security Challenges and Solutions: A Review by OSI Layers" S. SULLIVAN1, ALESSANDRO BRIGHENTE 2, (Student Member, IEEE), S. A. P. KUMAR 3, (Senior Member, IEEE), AND M. CONTI 2, (Senior Member, IEEE)
43. 3GPP TS 33.501 Release 15, June 2018

44. H. Fang, X. Wang, and S. Tomasin, "Machine Learning for Intelligent Authentication in 5G and Beyond Wireless Networks," IEEE Wireless Communications, vol. 26, no. 5, pp. 55–61, Oct. 2019.
45. ENISA THREAT LANDSCAPE FOR 5G NETWORKS, November 2019
46. AI for Beyond 5G Networks: A Cyber-Security Defense or Offense Enabler? Chafika Benza¨ıd∗ and Tarik Taleb† ∗†Aalto University, Espoo, Finland †University of Oulu, Oulu, Finland †Sejong University, Seoul, South Korea

Index

Note: - Page numbers with *Italics* for figures and **bold** for tables.

Printed in the United States
by Baker & Taylor Publisher Services